Multimedia
Networking

Other McGraw-Hill Books of Related Interest

Multimedia Networking

Bohdan O. Szuprowicz

McGraw-Hill, Inc.

New York   San Francisco   Washington, D.C.   Auckland   Bogotá
Caracas   Lisbon   London   Madrid   Mexico City   Milan
Montreal   New Delhi   San Juan   Singapore
Sydney   Tokyo   Toronto

Library of Congress Cataloging-in-Publication Data

Szuprowicz, Bohdan O.
    Multimedia networking / Bohdan O. Szuprowicz.
        p.   cm.
    Includes index.
    ISBN 0-07-063108-5 (pbk.)
    1. Computer networks.   2. Multimedia systems.   3. Telematics.
I. Title.
TK5105.6.S973   1995
650′.00285′66—dc20   94-38072
                                                                          CIP

1 2 3 4 5 6 7 8 9 0   DOC/DOC   9 0 0 9 8 7 6 5

ISBN 0-07-063108-5

The sponsoring editor for this book was Ron Powers, the editing supervisor was Jane Palmieri, and the production supervisor was Pamela Pelton. It was set in Century Schoolbook by McGraw-Hill's Professional Book Group composition unit.

Printed and bound by R. R. Donnelley & Sons Company.

McGraw-Hill books are available at special quantity discounts to use as premiums and sales promotions, or for use in corporate training programs. For more information, please write to the Director of Special Sales, McGraw-Hill, Inc., 11 West 19th Street, New York, NY 10011. Or contact your local bookstore.

To Totul

Contents

Introduction

The purpose of this book is to define and explain the ultimate objective of information technology, namely, interactive real-time multipoint multimedia communications. Networks with such characteristics are expected to provide future collaborative infrastructures for enhancing productivity and competitiveness of the interactive enterprise and the virtual corporation.

Interactive multimedia communications are the synthesis of TV concepts and computer capabilities into an integrated whole. This approach combines TV's power to grab and hold attention, motivate, excite, and indoctrinate, with the computing network's ability to randomly access, select, compare, transmit, and display data, information, and knowledge in form of text, graphics, images, animated sequences, voice, and video from a variety of sources within and without the enterprise.

Multimedia computing has been introduced at the standalone desktop level, and increasingly vendors of PCs, portable computers, and consumer compact-disk read-only-memory (CD-ROM) players are offering numerous multimedia-ready platforms at continuously declining prices. As a result, rapidly growing numbers of computer users familiar with multimedia computing are demanding connectivity with other multimedia users and sources across local and wide area networks (LANs and WANs) linking corporate workgroups and operating units throughout the world. The development of the digital information superhighway concepts and the plans of telephone companies and cable TV firms to supply interactive TV and other services to the home provide an additional impetus to the overall interest in interactive multimedia communications applications.

This book is designed specifically for information technology visionaries, executives, strategists, planners, managers, systems engineers, and developers who are facing the problem of introducing multimedia applications and communications into existing enterprisewide computing and networking environments.

The main focus of the book is identification of successful multimedia systems that enhance productivity and competitiveness of new restructured enterprises. It explores the potential of the information superhighway as a basis for new interactive marketing, scientific, and engineering services and entertainment businesses combining the elements of broadcasting, publishing, consumer electronics, telecommunications, and computing industries.

In the initial chapters of the book the reader will discover that interactive multimedia communications represent a group of key enabling technologies for restructuring the corporation and developing a more competitive posture on a worldwide basis.

The reader will quickly learn about the major strategic interactive multimedia applications implemented by the early adopters of these technologies. Whenever possible, such issues as business justification and design and implementation problems and solutions are clearly stated and described. Existing and future market opportunities are sized up and presented as potential development areas for immediate consideration.

A major section of the book is dedicated to the discussion of basic multimedia networking technologies. It provides the reader with a single source for understanding such diverse and critical issues as bandwidth, compression, standards, and networking architectures, as well as internetworking, client-servers, databases, and transmission facilities and their individual importance to the overall success of interactive multimedia implementations. This intelligence is complemented with discussions on multimedia networking hardware and specific issues involved in the development of multimedia networks, the information superhighway, and interactive TV.

The Promise of Multimedia Networking

The first chapter discusses the basic restructuring trends prevailing within contemporary corporate environments, which in many cases involve the interactive enterprise and virtual corporation concepts. Chapter 1 focuses on various levels of interactive multimedia computing and communications and their role in furthering the corporate restructuring processes toward the development of more productive and competitive enterprises of tomorrow.

The multimedia communications environment discussed in Chap. 2 is designed to explain the dynamics of interactive multimedia applications and the impact on various end-user audiences. It looks at different user groups and the basics of collaborative processes as well as delivery methods and platforms. The reader is made aware of the existence of various factors that influence these processes and are not always obvious to professional technologists. These factors include specifics of image-based conversational environments; problems associated with simultaneous use of voice, data, text, image, and video; integration of networking infrastructures; user interfaces; and real-time response issues.

Chapter 3 identifies several major industries that stand to gain from multimedia technologies and are having an impact on their development. It focuses on the emergence of new multimedia applications and businesses which are being developed through strategic alliances of existing enterprises in telecommunications, broadcasting, entertainment, publishing, and computer industries. The rush to acquire a position in what is being perceived as the massive multi-billion-dollar markets of tomorrow has an accelerating effect on investment in new multimedia technologies, products, and ventures. This process, in turn, benefits all who are faced with the need to plan interactive multimedia networks, and this chapter reminds them to keep their eyes open for new opportunities as new products come to market at rapidly falling prices. It also warns the reader about the rapid obsolescence factors that exist in various product lines and helps the reader understand the overall progress of the interactive multimedia communications phenomenon.

Strategic Multimedia Implementations

Starting with Chap. 4, the book provides the reader with detailed discussions of major networked multimedia applications. The first of these is multimedia videoconferencing, which is considered by many analysts as the most promising and universal interactive multimedia networking application. All forms of videoconferencing are prime examples of such implementations ranging from relatively simple point-to-point videotelephony to very complex real-time multiuser, multipoint, group conferencing systems. The chapter covers all such videoconferencing concepts with examples of real applications and a list of vendors of associated hardware, software, and service suppliers. It provides the reader with a complete coverage of all the variants of videoconferencing uses and identifies sources of products and information that will facilitate the implementation of such projects.

Sales and marketing applications are the subject of Chap. 5. These are interactive multimedia applications that provide automation of sales forces and customer service networks. The reader will learn about portable multimedia applications, various forms of multimedia merchandising kiosks, public information services, and specific industry successes in travel, real estate, and financial services. This chapter also covers the advanced concepts of virtual reality and its application in the sales and marketing arena. As in other chapters descriptive of specific multimedia markets, the reader is provided with a list of vendors providing multimedia-related products and services for these markets.

Multimedia training and information networks are covered in Chap. 6. This type of application represents some of the earliest implementations of interactive multimedia technologies and is expected to constitute an increasingly important market as time progresses and more companies start operating as interactive enterprises moving toward becoming virtual corporate entities. Aside from various forms of training, this chapter also deals with corporate multimedia distribution networks and various surveillance, monitoring, and imaging systems that rely on multimedia networking of one form or another.

In Chap. 7 special attention is given to the health care industry applications. As a result of the everpresent need for visual demonstrations and group consultations over a wide area, this industry presents unique and compelling reasons for implementing interactive multimedia imaging and consultative networks. There are numerous factors conducive to such implementations, including high costs, current politics, better patient monitoring requirements, interchanges with medical insurance suppliers, continuous medical training, and frequent introduction of complex new drugs and therapies. The readers will be able to grasp the elements of the emerging field of telemedicine, which is based on interactive multimedia technologies.

Interactive multimedia consumer markets are covered in some detail in Chap. 8. The objective here is to provide the reader with some insight into various interactive multimedia networks that are being developed to distribute consumer information and entertainment services into the home. While these applications may not be of immediate value to business managers, their implementation often includes useful solutions of complex multimedia networking applications. The multi-billion-dollar scale of these markets also attracts significant investments and creates new ventures that develop concepts and products that are of value to all concerned. The reader should also keep in mind that many new interactive multimedia consumer networks will provide alternative distribution and marketing infrastructures for promotion of company products and services.

Multimedia Networking Technologies

In subsequent chapters the book looks at the specifics of multimedia transmission and networking technologies. Chapter 9 is among the most useful because it brings to the attention of the reader the basic multimedia networking problem. It defines bandwidth ranges required for transmission of specific multimedia content and compares those with bandwidth capacities of various digital transmission facilities and infrastructures. As a result, this chapter more or less outlines the basic issue confronting anyone who is responsible for implementation of networked multimedia applications. It provides a starting point and a yardstick to plan, measure, and design a multimedia project which will meet the stated objectives and can be accomplished at optimal cost.

A critical issue in all multimedia transmissions is data compression. This subject is covered in considerable detail in Chap. 10, which provides the reader with basic explanations of compression techniques, potential tradeoffs between speed and quality of multimedia transmissions, and alternative compression technologies. It also discusses the question of major compression standards such as Joint Photographic Experts Group (JPEG), Motion Picture Experts Group (MPEG), and H.261 and identifies leading suppliers of compression products.

Closely related to the issue of data compression is the question of transmission standards which enable wide area multimedia communications. Although this is a new and fluid arena, standards pertaining to all aspects of multimedia transmission are still being developed and implemented which, in turn, are an incentive to the manufacturers of multimedia communications equipment to develop new cost-effective products and services. Chapter 11 provides the reader with an explanation of all multimedia-related standards already in force or in development.

In Chap. 12 the reader will find a discussion of basic LAN architectures and their respective multimedia data handling capabilities. The chapter also outlines several solutions that are in use to enhance existing LAN architectures to handle multimedia traffic. Multimedia metropolitan area networks (MANs), WANs, and virtual LAN concepts are also discussed, and specialized vendors supplying multimedia LAN products are identified.

Interconnectivity of multimedia LANs and associated questions of time sensitivity of multimedia traffic are covered in Chap. 13. The reader is introduced to the issue of latency when considering multimedia transmissions across various networks and the promise of the asynchronous transfer mode (ATM) as an ideal multimedia networking solution.

Chapter 14 deals specifically with multimedia client-servers and databases and the requirements for massive storage to handle multimedia data. Different approaches to multimedia databases are discussed, and new specialized videoserver products and ventures are identified.

Multimedia transmission facilities in all their forms are the main subject of Chap. 15. This section is a detailed description of various transmission services and their relative capabilities for handling multimedia traffic. The much-touted digital superhighway, formally known as the *National Information Infrastructure* (NII), is also explained here together with multimedia strategies of telecommunications carriers.

Chapter 16 gives the reader a quick overview of the multimedia networking hardware products which form the basis of all the enabling technologies making interactive multimedia communications a practical reality. It covers specialized multimedia components such as digital signal processors (DSPs), codecs (coders-decoders), interface cards, bridges, routers, gateways, switches, and more specialized multimedia intelligent hubs and multipoint control units. As in other chapters in this category, major multimedia networking equipment vendors are identified and evaluated.

Special Features of the Book

The last chapter of the book departs from discussing multimedia networking technologies and provides the reader with a special overview of planning, design, management, and implementation issues specific to multimedia applications. It warns the reader about numerous nontechnical questions that must be addressed before undertaking a multimedia project of any type. These include cultural, artistic, aesthetic, legal, and licensing issues that are often absent in more conventional information processing and networking projects. Lack of understanding of these issues can result in poor quality, serious project delays, cost overruns, and even legal action, all of which can be avoided by making sure that all such questions are resolved before project implementation is begun.

The appendix of the book is another valuable section. It contains a specialized alphabetized list of suppliers of multimedia networking hardware, software, and services. The entries include company name, address, telephone, and fax (facsimile) numbers for all vendors identified at the end of individual chapters. As a result, the reader who is seeking information or a solution to a specific problem can quickly identify the major players in that specific market niche and will be in a position to take immediate action. This list, of course, is not complete because interactive multimedia communications is a field in considerable flux. New products and ventures enter the market every day, and the current list is offered primarily as a basis for immediate action and further research.

The book also includes an extensive glossary covering specific multimedia and networking terminology. It is designed for the reader who may not be familiar with either domain and is seeking quick and pertinent references on all aspects of multimedia networking.

Acknowledgments

This book is the result of a series of events that have originally been undertaken as independent business initiatives and activities by my 21st Century Research consultancy over several years. As a result I now come to realize that I am indebted to a number of people and organizations that unquestionably contributed to the writing of this book although at the time neither they nor I were aware of it.

I would like to express my greatest appreciation to Ed Wagner, president of Computer Technology Corporation in Charleston, South Carolina, who was the first to encourage my effort to prepare a series of reports on "Multimedia Technology," "IBM's Multimedia Policy," and "Multimedia Networking and Communications." The research leading to the preparation of these reports created the information base without which the writing of this book would not have been possible.

Special thanks must also go to Susan Garavaglia, who as vice president of Chase Manhattan Bank was the chairperson of the Society for Management of Artificial Intelligence Resources and Technology—Financial Services (SMART-F$). She facilitated the creation of the Multimedia Special Interest Group within the organization which I was privileged to chair in its initial years and where many issues pertinent to interactive multimedia communications have been discussed in great detail. These experiences provided invaluable insights into understanding the business multimedia user requirements.

I also want to express my appreciation to Steve Feher, conference manager for the Interactive Information (I^2) Expo. By inviting me to participate in the program committee, he provided me with an unusual opportunity to exchange views and ideas with numerous corporate multimedia communications practitioners.

There are many other people whom I want to thank for providing special opportunities to advance my understanding of interactive multimedia communications. Bill Atkins and Tom Veal of Deloitte & Touche are two partners of that firm who made it possible to test a number of interactive multimedia ideas at an early stage. Equally valuable were the observations and comments of many executives and employees of many corporate clients whose names escape me. They include people from companies such as Bristol-Myers-Squibb, Hoffmann-La Roche, Hoechst, New York Life, Prudential Insurance, Sandoz, and others. Some of the most valuable suggestions that I have incorporated in the writing of this book are an amalgam of comments and observations from real-life end users of multimedia systems in corporate environments.

I also want to thank several entities within the McGraw-Hill empire for perhaps indirectly and unwittingly assisting in the development of research and production of this book. The DATAPRO Information Services Group accepted many of my proposed reports of multimedia topics thereby creating another opportunity to develop additional insights into the subject and to evaluate numerous multimedia products and vendors. The *Data Communications* magazine deserves special thanks for giving permission to use many of their excellent figures that so well illustrate complex multimedia networking issues. I am equally grateful to *LAN Times* and *Byte* magazines for permission to use some of their figures to enhance the presentations in this book.

Last but not least by any means is the extraordinary patience and understanding of my wife Martyna who went out of her way all along to facilitate the writing and editing of this book.

Bohdan O. Szuprowicz

Multimedia
Networking

1

Restructuring the Enterprise

Economies of many industrialized countries around the world are going through a period of intensive restructuring resulting in new corporate operational models based on advanced information technologies to enhance their productivity and competitiveness.

The interactive enterprise concept is the key to these new business entities. It depends on development of collaborative multimedia computing at the individual, workgroup, and corporate levels and extensive use of client-server networking concepts.

Interactive enterprises representing various product lines, skills, and competencies can pool their resources to participate in virtual corporations organized to take advantage of specific market opportunities and more rapidly and efficiently than traditional enterprises.

Networked multimedia applications will become widely used in these new business environments, providing interactive real-time multiuser communications infrastructures. Computer hardware, software, and telecommunications vendors are now developing multimedia-enabled products and high-bandwidth networks to facilitate such operations.

Restructuring of the enterprise is a collective term which implies reengineering of corporate processes designed to make it more competitive in a rapidly changing global marketplace. New technologies and shortening product life cycles require the development of more efficient operations to survive and make a profit. Restructuring has been under way for some years in many industries and is expected to continue for some time to come, perhaps well into the next century.

With more powerful hardware at the desktop and in the field being linked within networks, operating groups and business units will develop into interactive working environments offering new efficiencies and competitive advantages to corporations with sufficient vision to take the necessary action today.

Interactive multimedia communications are the basis of these new solutions, creating the real-time interactive infrastructures that will make virtual corporate model a reality. This is the new corporate paradigm for the 1990s which is expected to break down the barriers of time and distance and create more effective collaboration between all participating business units of a virtual corporate entity.

The Virtual Corporation

The concept of the virtual corporation has been advanced by leading management consultants during the last few years. It has been outlined in some detail in *The Virtual Corporation: Structuring and Revitalizing the Corporation of the 21st Century* by William H. Davidow and Michael S. Malone (HarperBusiness, 1992). Recent combinations of companies like General Instruments, Intel, and Microsoft to exploit the interactive TV business opportunities are examples of a virtual corporate entity in the making. According to market research organizations such as The Yankee Group (Boston) or Ovum Ltd., rapid expansion of mobile technologies is also facilitating and accelerating the emergence of the virtual corporation.

Traditionally corporations are entities that are tied to a production base and survive and prosper by figuring out how to sell their products to what customers at prices that will result in maximum profits. The virtual corporation concept reverses that process. It mandates that companies continue to discover what customers want or need and instantly respond with whatever action is necessary to meet those customer demands. Executives of a virtual corporation are not running static organizations but focus on achieving specific business results in terms of customer satisfaction.

In an increasingly competitive global economy, customer satisfaction is indeed seen as the key to future survival. In fact, customer satisfaction is the basic entry point at which competition now begins. In order to capture a leading market share and maintain it for any length of time, companies must go beyond customer satisfaction and move toward customers' delight. The objective is to discover how to deliver products or services that satisfy customers so completely and continuously that they will stay with you through all your product generations (see Table 1.1).

To achieve such goals a company must be extremely flexible. It must have exquisite vision, superb market intelligence, and the ability to conceive, design, test, manufacture, and deliver customized products faster than any competitor anywhere in the world.

A *virtual corporation* is a temporary network of independent operating units such as creative designers, manufacturers, suppliers, customers, and other experts in marketing and finance linked by interactive multimedia networks to share skills, costs, production facilities, resources, and access to each other's markets.

TABLE 1.1   Characteristics of Traditional and Virtual Corporate Models

Characteristics	Traditional	Virtual
Focus	Product service	Customer satisfaction
Operation	Existing and predictable products and outlets and delivery time defined by the corporation	Creation and delivery of any product, any time and at any place to satisfy immediate customer demands

It is a highly flexible organism consisting of various collaborative units that pool their resources to exploit a specific and timely business opportunity that one of the units discovered as a result of superior or more timely market intelligence. When such an opportunity is exploited and competition catches on to the opportunity, the virtual corporation ceases its operations and may disappear forever. The concept is illustrated in Fig. 1.1.

This is in contrast to the typical joint ventures and alliances, most of which are formed on the assumption of an ongoing relationship to exploit market opportunities that are not necessarily clearly defined or specified. The virtual

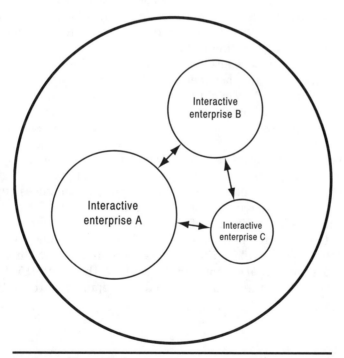

Figure 1.1 The virtual corporation concept. (*Source: 21st Century Research.*)

corporation, also known as the *modular corporation,* is a new corporate concept that is specifically designed to exploit market opportunities that are increasingly fast-changing events often limited to a single product line. Such markets are characterized by rapid growth rates and very short product life cycles. Under the circumstances speed and time to market become more critical than other factors for success. The virtual corporation must move rapidly in order to make a profit before competitors acquire a market share so large that it is not worth remaining in that business any more.

Conventional, vertically integrated business structures are not able to respond rapidly enough to meet such marketing challenges. The virtual corporation model, however, allows concentration on the specific core activities of an enterprise and takes advantage of outside specialties to develop, or manufacture other aspects of a product where they have a competitive advantage as a result of special skills, patents, or competencies. This is particularly the case in technology-oriented industries such as electronics or pharmaceuticals but is also true in other areas such as financial services.

The virtual corporation implies a competitive advantage or even survival in an increasingly interdependent global economy. It represents the ability to develop and deliver a real product or service any time, any place, in any quantity but also faster than any competitor. Associated with this approach is also the need for a drastic change in the mindset of business executives and workers of traditional enterprises. They must switch their marketing perspectives and work habits from mass production to mass customization of goods where customer service is the key driving factor.

The virtual corporation concept depends on advanced information technology to facilitate the interaction between executives and workers on a real-time basis. It also draws on an array of dynamic integrated databases which contain the latest information and knowledge about markets and customers and their latest needs, technologies, products, designs, manufacturing suppliers, and competitive threats.

All this is part of the business revolution which is already under way as companies engage in reengineering their organizations and processes, trying to become more flexible and exploit rapidly developing new business opportunities. Companies that fail to reorganize to meet the new era are facing very stiff competition and possibly extinction if they do not move fast enough. A lot of this reorganization is taking place in the United States, and business analysts believe that if this process will continue across all industries, the United States may become a leading virtual economy in the world by the year 2015. The alternative is to become the largest postindustrial developing country of the third world.

The Interactive Enterprise

Corporate organizations rely extensively on large numbers of different computing and communication devices and systems. These range from notebooks and personal digital assistants (PDAs) in the field, through desktops and workstations interconnected through LANs with minicomputers and mainframes located in many parts of the country or across the world. For a virtual corporation to become an effective reality, such systems must be able to exchange everything from simple data files to high-resolution visualizations, and multimedia presentations that may include real-time interactive multiparty communications. The necessity to interconnect new and unfamiliar enterprises, their customers, and other supporting organizations is an enormous challenge to business entities of tomorrow. Such interactive enterprises may consist of highly specialized independent businesses or specific business units of large corporations with considerable freedom of action in the marketplace and their own informational infrastructures. Figure 1.2 illustrates the concept of the interactive enterprise and the associated virtual conferencing idea.

If an organization wants to participate in a virtual corporate operation, it must develop within its environment interactive multimedia computing and communications infrastructures. Their networks must be able to integrate easily to allow collaborative multimedia computing at the workgroup, corporate, and intercompany levels on a global basis. Telecollaboration is now seen as a powerful means of communications within the interactive enterprise of the future and provides the real technological basis for the virtual corporation operational and marketing concept.

To be considered an interactive enterprise, a business unit must meet a number of requirements. It must possess a flexible communications center linking each individual to the enterprise, integrating critical office productivity tools on the desktops, and providing full multimedia transmission capabilities. This means very high-performance workstations or client servers, high-bandwidth networking capacity, multitasking, and cross-platform standards that allow communications with remote locations and mobile units.

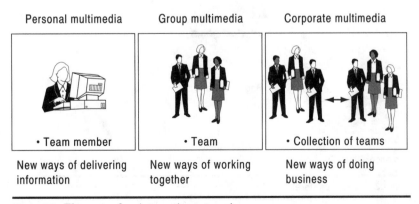

Figure 1.2 Elements of an interactive enterprise.

Collaborative Multimedia Computing

The interactive enterprise must streamline its operations to eliminate redundant and inefficient activities and provide facilities for collaboration of any corporate worker with anyone, anywhere in the world, regardless of time, location, computer hardware and software deployed, or actual differences. This means the use of local area networks (LANs), metropolitan area networks (MANs), wide area networks (WANs), and internetworking facilities that can handle massive, multiuser, real-time, and interactive multimedia data traffic with relatively low latency. This also means multimedia data handling capabilities within databases, client-servers, bridges, routers, gateways, and on LANs and backbone networks with real-time video transmission capabilities in all the interconnections.

The Virtual Conferencing Idea

Real-time audio and video transmission capabilities make it possible to conceive of virtual conferencing rooms or environments that provide linkage at the most convenient moment in time with key representatives from marketing, research, development, outside vendors, distributors, and customers. These individuals can collaborate on a project without leaving their places of work where they have ready access to all the most pertinent data and information (Fig. 1.3).

Behavioral research suggests that for most effective conferencing no more than six participants at any one time should be involved. Many multimedia conferencing systems today are designed with this limit in mind. On the other

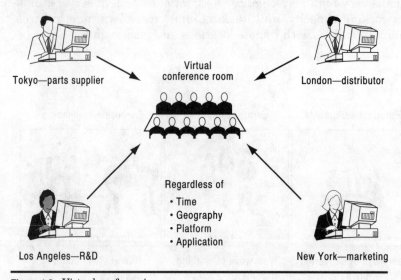

Tokyo—parts supplier

Virtual conference room

London—distributor

Regardless of
• Time
• Geography
• Platform
• Application

Los Angeles—R&D

New York—marketing

Figure 1.3 Virtual conferencing.

hand, several virtual conferences can be taking place simultaneously within an interactive enterprise that may also be involved in multiple virtual corporate projects. As a result, multimedia transmission facilities above the workgroup level should have the capacity to handle numerous simultaneous multimedia transmissions.

The capacity of such links is not readily predictable because it depends on the specifics of each virtual corporate effort, which vary in time and scope. This presents an opportunity for individual multimedia services that may develop to provide such facilities on a temporary basis to groups of interactive enterprises combining into virtual corporate organizations.

Role of Standalone Multimedia Platforms

Desktop multimedia platforms provide a new way of delivering information in any way an individual wants to look at it. Originally these were one-way communications tools designed to provide corporate training, sales materials, customer service, and some just-in-time diagnostics and factory floor support. Basic elements of a personal multimedia platform are illustrated in Fig. 1.4.

Personal multimedia facilities consist mostly of standalone PCs and workstations using CD-ROM drives and specialized software for handling full-

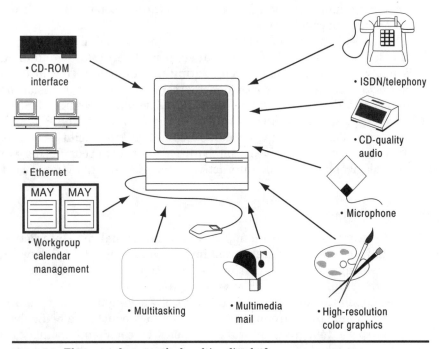

Figure 1.4 Elements of a networked multimedia platform.

motion video. Most applications consist of interactive programs prepared by third parties providing information, training, or specialized presentations. Most personal multimedia platforms are basically one-way communications systems, but an increasing number of such systems are linked to LANs and even WANs. Some multimedia platforms provide access to object databases (ODBs) and other files that can hold massive binary large objects (BLOBs) of multimedia data. Multimedia products are mostly distributed in form of CD-ROMs or laser disks on which interactive multimedia applications have been developed by third-party contractors. Most networks do not provide sufficient bandwidth for real-time multimedia transmissions that are required for multimedia conferencing and communications within an interactive enterprise to be effective for engaging in the virtual corporation type of activity.

An increasing number of PCs are already connected to a LAN, and as collaborative computing becomes more widespread, multimedia-capable PCs on networks will be more common. This results in data transmission overloads and bottlenecks on the networks intolerable by time-sensitive transmissions. This, in turn, calls for communication links that are faster and offer much more bandwidth capacity than existing facilities.

IBM estimated recently that by 1995 20 million PCs will be connected to some network. If only 15 percent of those are multimedia-capable platforms, this represents several million PCs transmitting and manipulating massive volumes of multimedia data. This percentage may be even higher because an increasing number of new PCs and workstations come with built-in multimedia processing capabilities.

Vendors are quite aware of these developments and are manufacturing multimedia hardware and software that support DOS, OS/2, Windows (Microsoft), Windows NT, Apple, and UNIX operating environments. They also provide new products facilitating low-cost desktop conferencing, compression devices for audio and video storage, multimedia digital signal processing (DSP) chips, and video distribution systems. This proliferation of multimedia PCs and workstations as well as upgrading to high-speed networking facilities create an infrastructure that is conducive to collaborative multimedia computing, which is the technological basis for the interactive enterprise and the virtual corporation.

Group Multimedia Networks

The term *group multimedia* implies at the outset real-time interactive facilities for sharing screens, whiteboards, images, and other data files within a teleconferencing environment. It is the first step toward the development of collaborative multimedia work environments.

Group multimedia provides new ways of working together on a project team basis and numerous groupware concepts and products are being developed for this purpose. In a group multimedia operation each team member is equipped with a networked PC or workstation which provides multiple-way communications between team members. Theoretically group multimedia systems provide the interaction with project team members who have different skills and other viewpoints that contribute to a more innovative atmosphere resulting in better products or solutions.

Typical group multimedia activities include collaborative business analysis and decision making, concurrent or simultaneous engineering, and desktop telephony, including E-mail, faxes, voicemail, and specialized graphical user interface (GUI) interfaces for managing such resources.

Group multimedia systems deliver an interactive communications infrastructure for many financial service organizations whose branches may be geographically dispersed but must work together reviewing and analyzing compound documents consisting of forms, reports, images, voice annotations, video clips, faxes, phone messages, and database records. When videocameras and real-time transmission facilities are included, group multimedia systems provide multiuser conferencing and become components of enterprisewide multimedia infrastructure that makes the interactive enterprise possible.

Corporate Multimedia Systems

Desktop standalone multimedia PCs deliver information in a better way, but group multimedia is a new way of working together. By comparison, corporate multimedia means a new way of doing business. It is the ultimate collaborative multimedia concept within a corporation and contributes to more effective interaction between various corporate groups in discovering better ways of developing new products and providing more competitive services. Some companies with vision realize that advances in information technology provide new opportunities for enhancing their operations, but changes are required in their management structures and decision processes to put such systems into effect. Corporate multimedia provides the infrastructure that dynamically facilitates such transformations. (See Chap. 12 for a complete discussion of multimedia LAN alternatives.)

Corporate multimedia empowers workgroups throughout the company. It provides information which results in a competitive advantage through enhanced and more effective communications. This results in faster decision-making processes, better customer services, higher-quality products, more creativity, and increased group productivity.

The benefits of corporate multimedia will not take place until an appropriate networking infrastructure is in place, but they also depend on the content of multimedia applications and their form of presentation to the various end-user populations. Technology alone is not sufficient to make corporate multimedia a successful alternative.

Networked Multimedia Applications

Networked multimedia applications are designed to make information technology central to solving business problems. Multimedia conferencing, which provides an all-purpose communications facility, is the best example of such an implementation but in and of itself it is of no value.

Networked multimedia applications support corporate communications, marketing, sales, training, product development, manufacturing, and administration. Table 1.2 illustrates the different categories and indicates more details about specific multimedia applications within those corporate function areas.

Until recently there were few products facilitating multimedia communications on corporate LANs and WANs. This situation is changing quite rapidly, and many of the existing multimedia applications are prime candidates for upgrading to networked solutions status.

TABLE 1.2   Networked Multimedia Implementations of Corporate Functions

Function	Networked multimedia implementation
Administration	Multimedia databases
	Executive information systems (EISs)
	Multimedia file sharing
	Document imaging
	Productivity systems
Communications	Multimedia conferencing
	Corporate broadcasting
	Videomail systems
	Employee information systems
Manufacturing	Testing procedures
	Equipment maintenance
	Diagnostics
Marketing	Interactive advertising
	Product information kiosks
	Promotional presentations
	Customer services
	Virtual reality simulations
Product design	Concurrent engineering
	Product simulations
	Visualizations
	Design advisory systems
	Product references
Sales	Product catalogs
	Merchandising kiosks
	Buyers' workstations
	Portable multimedia presentations
	Selling advisors
	Personal digital assistants
Training	Corporate training
	Management training
	Sales training
	Distant training
	Just-in-time training

Intelligent Multimedia Networks

Numerous multimedia implementations are similar to expert systems, particularly in diagnostics, customer support, help desks, and training applications. Many expert systems that were originally developed with only text and rudimentary graphics in their knowledge bases can be enhanced significantly with multimedia objects and elements.

In fact, most recent expert systems development tools are now capable of handling objects which may include graphics, images, animation, and audio and video sequences. On the other hand, multimedia applications are most often developed with authoring systems which are examples of particularly user-friendly object-handling development tools. Some of those tools now include features such as an inferencing engine, extensive branching, and performance scoring which can provide significant levels of artificial intelligence in multimedia applications.

Expert systems and authoring systems can both be used to develop an intelligent interactive multimedia system and seem to fit rather well in such environments. Within an interactive enterprise network, expert systems should also reside in readiness to provide immediate guidance and advice to anyone of the workforce who will often be required to make quick decisions about relatively new and unfamiliar issues. Corporate intelligence would be responsible for finding and obtaining the appropriate expert systems to perform such tasks.

Implications of the Digital Superhighway

Within the corporate environments networked multimedia use private LANs and WANs and special long-distance carrier facilities, but this situation is changing rapidly. Major telecommunications firms, cable TV companies, and entertainment organizations are contemplating a much larger multimedia market. There are 92 million American households most equipped with televisions and telephones, 25 million of which use personal computers. Here they see a market into which they want to inject interactive multimedia services in form of 500 interactive TV channels that will provide consumers with entertainment on demand, interactive home shopping, promotion, advertising, and interactive multiuser games.

Such developments will automatically create numerous new business opportunities. These will be readily exploited by interactive enterprises and virtual corporations because such organizations will have the capabilities to interconnect with other multimedia networks to satisfy the fleeting demands of even small groups of customers at a profit before other competitors can even design a product for such markets. Figure 1.5 illustrates the potential infrastructure and numerous access points to the digital superhighway of tomorrow.

Some elements of the information superhighway are already in place. Internet, while primarily handling text and data, is sometimes seen as a prototype of a network of networks and already links some 20 million people and over $1\frac{1}{2}$ million computers throughout the world. Some innovative vendors are already developing software to handle multimedia transmissions along the Internet and enhance its bandwidth by providing cable TV transport facilities. For the most part, however, it is most useful as a global E-mail system, although it is still relatively difficult to use.

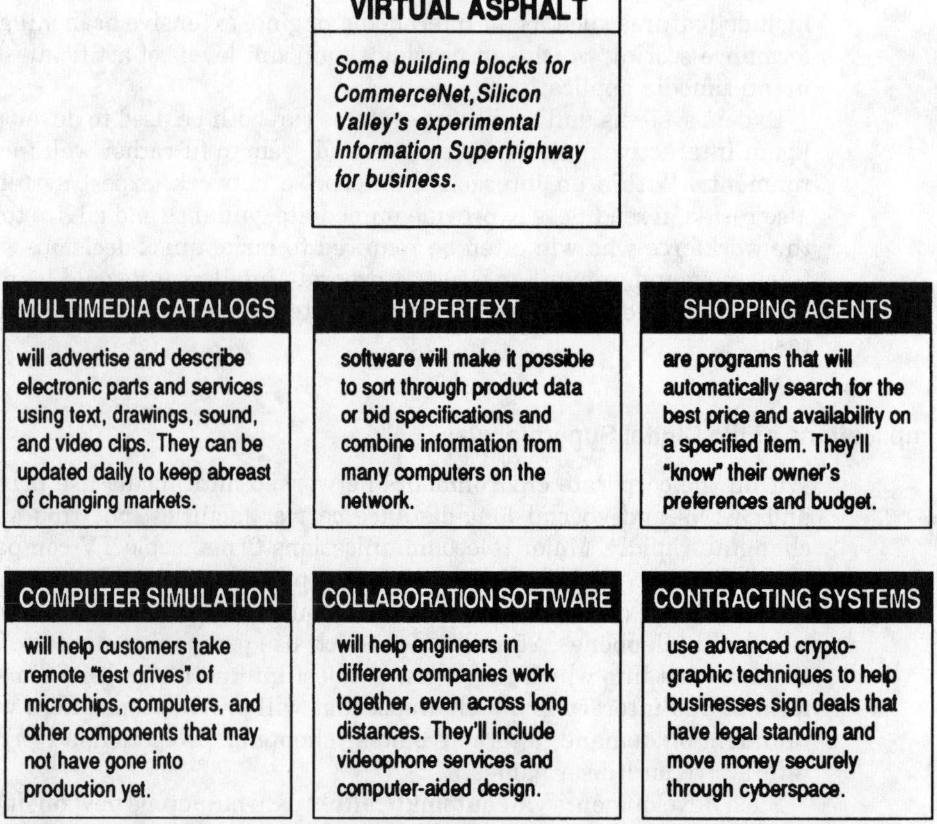

VIRTUAL ASPHALT

Some building blocks for CommerceNet, Silicon Valley's experimental Information Superhighway for business.

MULTIMEDIA CATALOGS

will advertise and describe electronic parts and services using text, drawings, sound, and video clips. They can be updated daily to keep abreast of changing markets.

HYPERTEXT

software will make it possible to sort through product data or bid specifications and combine information from many computers on the network.

SHOPPING AGENTS

are programs that will automatically search for the best price and availability on a specified item. They'll "know" their owner's preferences and budget.

COMPUTER SIMULATION

will help customers take remote "test drives" of microchips, computers, and other components that may not have gone into production yet.

COLLABORATION SOFTWARE

will help engineers in different companies work together, even across long distances. They'll include videophone services and computer-aided design.

CONTRACTING SYSTEMS

use advanced cryptographic techniques to help businesses sign deals that have legal standing and move money securely through cyberspace.

Figure 1.5 Digital superhighway potential. (*Source: Reprinted from April 18, 1994, issue of* Business Week *by special permission, copyright © 1994 by McGraw-Hill, Inc.*)

Interactive Multimedia Communications Basics

A number of factors must converge in order to justify the development and operation of an interactive multimedia communications network in a business environment. Businesses are becoming fragmented and decentralized in order to better meet customer preferences, which vary increasingly according to geography, demographics, and economic development levels in a region. An average Fortune 1000 company in the United States already operates about 254 sites that are widely dispersed and 54 percent of which employ fewer than 20 workers. This type of corporate environment reflects the fragmentation of social and business fabrics in modern societies in large measure resulting from widespread information availability and dissemination. This situation also sets the scene for wide area interactive multimedia communications as a most attractive alternative of corporate communications and marketing activities.

Under the circumstances it is necessary to take into account a number of factors such as interactivity levels; modes of conversation; the nature of the collaborative process; existing platforms and systems; simultaneous use of voice, text, image, data, and video; integration of networking infrastructures; end-user interfaces; and real-time response requirements (Table 2.1).

Success of an interactive multimedia communications system will also depend on the system's ability to handle multimedia elements simultaneously through an integrated real-time networking infrastructure that puts the minimal demands on the end user as far as computer literacy and technology know-how are concerned. At the same time it must create a close simulation of real-life conversations including unrestricted exchange of ideas, data, images, and documents.

TABLE 2.1   **Critical Factors Affecting Multimedia Communications**

Major factor	Description and comments
Interactivity levels	Definition of relationships of user to documents, platforms, and other users with regard to access, distribution, and manipulation of multimedia content
Conversation modes	Defines the nature of interactive conversations in terms of multimedia elements manipulated on screens
Collaborative process	Determines the numbers and locations of users involved in the collaborative process in terms of their proximity to each other, frequency of conversation, and need to communicate
Existing platforms	Describes the existing and planned information technology hardware and software infrastructure and its capability to support interactive multimedia communications
Multimedia content	Determines the type, volume, and sources of multimedia elements such as text, data, voice, images, sound, and video that must be integrated in the interactive system
Networking integration	Assessment of the existing networking infrastructures and their capability to support multimedia communications bandwidths and connectivity with other networks within and without the corporation
User interfaces	Determines the existence and suitability of user interfaces such as GUIs that are optimal for interactive multimedia communications
Real-time response	Defines which segments of interactive multimedia communications must support time-sensitive multimedia transmissions in real time to make their introduction and application a practical proposition

Levels of Interactivity

In order to understand all the issues and problems associated with interactive multimedia networking and communications it is necessary to define and classify the various levels and categories of interactivity that come into play.

Interactivity is best defined by the type of multimedia information flows. These are broadly categorized into traditional transactions between a user and specific documents, more exploratory interactions between a user and various delivery platforms, and collaborative transactions between two or more users. Within each of these categories multimedia information flows depend on interactive access interfaces, broadcasting facilities, or object-oriented manipulation of unstructured multimedia elements (Fig. 2.1).

In all cases such interactivity requires transmission facilities or networks with sufficient bandwidth to satisfy specific end-user expectations. However, interactivity requirements of end users who are mostly involved in local information or knowledge retrieval are relatively modest when compared with real-time multipoint conferencing involving groups of end users widely scattered

Multimedia applications can be viewed in terms of different communications models and varying information flows.

	User-to-documents	User-to-computer	User-to-user
Object-oriented manipulation	Mail	Database	Groupware
Broadcast	Newsletter	Information kiosk	Presentation
Interactive access	Hypermedia	Graphical user interface	Conferencing, training

Figure 2.1 Interactive multimedia information flows. (*Source: Reprinted from* Data Communications, *September 1992, p. 87, copyright by McGraw-Hill Inc., all rights reserved.*)

throughout the world and in different time zones. As a result, bandwidth requirements of the different networking segments within and without an interactive enterprise will vary widely and must be interfaced to operate in a smooth and seamless manner whenever possible.

End-user–documents interactivity

The interactivity between users and documents at its simplest is exemplified by hypermedia applications. These are hypertext products that can also manipulate certain graphic, image, and video elements. Many of these are based on CD-ROMs and standalone multimedia platforms, but increasingly specialized corporate hypermedia applications may require client-server or other storage systems where updating can take place by persons other than the end user. In the broadcasting mode corporate videos are prepared for information purposes and made available through networks to desktop users. Interactivity here is limited to access by end users at times most convenient to them, and often information may be downloaded into local storage for future viewing. Multimedia versions of E-mail also fall under this category, where users can access messages in specific locations as well as create their own messages for transmission to others in predetermined locations. In all these cases users are basically unable or severely restricted in the ability to manipulate or change the multimedia contents of these transmissions.

End-user–computer interactivity

This type of interactivity is more sophisticated and characterized by GUIs that provide the user with numerous choices of action. Basically such GUIs allow interactive access to a wide variety of tools that can be used for development of multimedia applications. Most authoring systems are excellent examples of this genre of software. Characteristically this type of interactivity is not time-sensitive in and of itself, although it includes software tools that can be used to develop and store such multimedia applications.

In the broadcasting mode this type of interactivity is best represented by public or private kiosk-based applications. These may range from relatively simple information dispensers to sophisticated merchandising systems with considerable built-in intelligence and order-taking capabilities. In most instances such interactive multimedia kiosks are designed around a specialized server and use LANs or WANs for distribution and updating of multimedia information flows.

Multimedia databases, multimedia data files, and ODBs capable of handling multimedia or binary large BLOBs represent various multimedia storage facilities. These are accessed by users and developers who seek multimedia elements for use in specific applications. Such objects can be manipulated by changing their characteristics such as color, size, texture, or shape. Through a combination of various software tools accessible by way of the GUIs, users and developers can conceive and design original multimedia objects as well as capture existing images, sounds, or videos for storage and manipulation at a later date.

Interactivity among end users

This form of interactivity is best represented by multimedia conferencing, and its main characteristic is the necessity to operate in real time. As a result, all multimedia applications in this category are time-sensitive and as such impose extreme bandwidth and speed requirements on associated networks and transaction facilities.

Multimedia conferencing networks provide direct communications between two or more parties who expect to use such facilities not unlike the conventional telephone networks. Multimedia conferencing systems will vary in complexity depending on whether these are point-to-point, person-to-person, multipoint, multiuser, and multiprotocol implementations. Such interactivity may represent *virtual conferencing* between parties, or corporate training and *just-in-time* instructional sessions, but in all cases these are real-time interchanges requiring considerable bandwidth capacities and high-performance equipment to operate.

Interactive multimedia business presentations are often special events where one group of users controls the broadcast of multimedia information to larger audiences. In such cases the objective is to discover group preferences or establish consensus through real-time feedback and manipulation of the presentation itself. Interactivity of this type may require special presentation facilities with large screens and audience response units. It may also involve the use of networks or wireless communication facilities.

Telecollaboration or groupware is the most sophisticated and complex form of user-to-user interactivity. It involves multimedia conferencing as well as simultaneous real-time capabilities for manipulating objects, images, drawings, text, and videos within a group. This type of application requires extensive multiuser communications bandwidth capacity as well as simultaneous access to various multimedia databases and storage systems. Concurrent or simultaneous engineering systems used in some manufacturing industries are good examples of such group interactivity.

Multimedia Communications Elements

The requirements for interactive multimedia networks will vary widely from application to application within and without an organization. The final selection of all the components in such a system will be governed by the needs of the targeted audiences, but certain key factors are basic to justifiable multimedia communications systems. It must also be understood that such factors are continuously affected by rapidly changing prices of associated hardware and software and the introduction of more innovative technologies.

The very basic requirement for establishing an interactive multimedia communications system or network of any kind in a working environment is the existence of human interaction. A recent study by NYNEX in support of their development work on Media Broadband Services (MBS) recognizes many communities of interest within an organization which are established and maintained primarily by various forms of conversation. Such interactions between individuals and groups usually develop into networks of relationships pursuing a common or mutually advantageous objective and result in collaboration and accomplishment of work.

Conversation as such implies a number of fundamental principles which make it useful and should be emulated in a collaborative multimedia work environment to make it a success. These principles include expression, articulation, response, exchange, dialog, and improvisation (Table 2.2). These should not be restricted by sets of often awkward and limiting procedures as in traditional computing environments. All the preceding principles are necessary factors for the design and implementation of an acceptable collaborative multimedia environment, but this does not mean that they are sufficient to justify such a network. Economic factors will ultimately prevail, but it is important to keep in mind that a justifiable system that does not reflect actual conversational realities and practice of a given environment is often doomed to failure.

Where interactive multimedia communications are concerned it is also necessary to establish at the outset that collaboration between individuals and groups involves a relatively significant amount of image-based conversations. This may be readily ascertained by studying the existing conversational

TABLE 2.2 Fundamental Principles of Conversation

Principle	Definition and system equivalent
Expression	Process of representing ideas and concepts in words and through other media such as images and models; system equivalents include monitors, screens, colors, audio, and video
Articulation	Action of bringing together various elements to create overall image or idea; performed by GUIs, authoring systems and specialized interactive presentation software
Response	An act of independent reaction to expressions and articulations; equivalents are telephones, information booths, kiosks, and videoconferencing systems
Exchange	Transfer of ideas or information in consideration of those received; system equivalents include shared screens, whiteboards, annotations, and E-mail systems
Dialog	An exchange of ideas or opinions between two or more parties simultaneously; multiparty teleconferencing and MCUs are system equivalents and enabling technologies
Improvisation	Ability to make, invent, or arrange an unforeseen expression on the spur of the moment; real-time transmission using graphics, scanners, and camcorders is best equivalent

exchanges between parties involved in form of conferences, presentations, images, drawings, graphics, sketches, photographs, videos, and films as well as inspection visits to distant locations.

It is also necessary to take into account certain quantitative and legacy factors pertaining to existing communications infrastructures. Legacy systems include mainframe-based information systems that are critical to running day-to-day operations of major corporations. It is necessary to determine to what extent these can be further exploited and what additional investments will be required to develop a multimedia communications environment that will satisfy the demands of all conversational communities within an enterprise.

If a virtual corporation operational model is being pursued, considerable flexibility must be built into the conceptual design. It will be necessary to accommodate the temporary collaborations of unpredictable interactive groupings and enterprises within and without an organization.

The Collaborative Process

The term *collaborative multimedia process* implies that numerous workers are always engaged in conversations with each other and that they are dispersed at different locations. Nevertheless, their proximity to each other, need to communicate, and frequency of conversations have a bearing on whether an interactive multimedia system is justifiable. It is very hard to justify such a system for a group of workers who are located next to each other or who can readily converse with each other by getting together in a nearby conference room.

On the other hand, if there is a lot of telephone and fax traffic, transmission of data and documents, as well as travel involved in carrying on such collaboration and the groups of collaborating workers exist in distant cities or even countries, these are clear indications for establishing multimedia conference facilities of collaboration. It is difficult to determine how many collaborating workers are needed to justify a multimedia network of this type because the final decision is influenced by the length and frequency of conversations, volume of information, images, and data exchanged. Nevertheless multimedia conferencing of this category has been implemented by major corporations in cases where relatively few nodes of interaction were in existence.

One of the most important aspects of an interactive multiuser multimedia communications network linking several participants is the requirement for several simultaneous transmission channels. Each station in such a network must have the capability of generating one outgoing channel of multimedia audio and video as well as accepting and decoding several other channels originating from all the other collaborating stations. In the final analysis this means significant investments in upgrading existing LAN and WAN systems for processing multiple streams of multimedia data in a real-time environment. In practice, participating stations should be able to conduct unrestricted multimedia conferencing with no more than about 6 to 10 other stations. Various studies suggest that conversations within groups larger than that become very hard to manage and are counterproductive (Fig. 2.2).

Some early adopters have already implemented multimedia conferencing systems of relatively few nodes. The Advanced Technology Group at Apple Computer began using 12 nodes in a network, while their corporate business unit has one consisting of 15 nodes. Another early adopter is Motorola, where a 30-node system is in use within a business communications group. A much larger network exists at Aetna Life which links many branches with over 100 nodes in its multimedia network. Other early adopters that already introduced multimedia conferencing networks include the Federal Court System, Ford, Hewlett-Packard, Kaiser, and Levi Strauss.

In design of corporate multimedia networks of this type, it is also necessary to define and distinguish different potential user groups. Most are company employees but clearly at different responsibility levels. This suggests the potential for discrete networks depending on corporate function and responsibility levels. These may link in-office employees with field sales forces and even telecommuters working in their homes or consultants and contractors equipped with appropriate platforms, modems, and passwords. One important consideration in such networks is the question of data security and access, which must be carefully designed to make sure that intermixing of data and inputs from numerous sources in real time does not threaten corporate confidentiality.

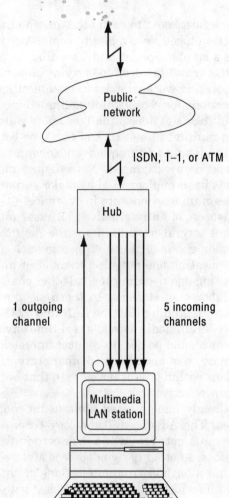

Figure 2.2 Multiple channels in multimedia conferencing.

Special consideration must be given to nonparticipatory user categories that may want random access to multimedia communications environments. Such users include observers at higher management levels and trainees or other persons who need to be informed about the occurrence of a particular conference and its results either in real time or after the conclusion of a conversation but who are not active participants in such conversations.

Multiple Platforms and Processes

The most obvious multimedia communications network that comes to mind is a group of identical PCs or workstations on a particular LAN, all of which have almost identical characteristics and capabilities. While this may sometimes be the case within a small and isolated LAN environment, once several groups are involved in distant locations or across international borders, the picture changes drastically.

The most likely multimedia communications systems will have to be imposed on existing multiple processes and platforms. This is true for personal hardware and software, as well as for networks and transmission facilities. As a result, the overall system must be designed with the most demanding process in mind. In the multimedia communications world this will most likely be the interactive real-time multiuser full-motion video conferencing mode.

On the other hand, this does not mean that all transactions on such a network will take place in such a mode all the time. Nor does it mean that all existing stations within the system could operate in such a mode simultaneously. Existing informational infrastructures, technological, and budgetary limitations will come into play at least in the initial stages of implementation. Nevertheless, it is important to keep in mind that between 6 and 10 participants should be able to conduct an unrestricted multimedia conference at any one time to make these environments viable.

Depending on the nature of conversations and conferences under way, different transaction processes will come into play. As an example, a multimedia conference will require interactive real-time video transmission via LANs and WANs and private or public high-speed networks but a multimedia training session which may not be rated as time-sensitive will need only LAN access to a video or a file server at best. Similarly, multimedia applications development at one of the stations may take place almost exclusively at the desktop except for brief network transmissions to download video, audio, or data segments from specific corporate databases. As far as corporate broadcasting is concerned, the process becomes a one-way video distribution service to the employees and although its frequency is unpredictable, it is short-lived and can be scheduled well in advance as well as downloaded for future viewing. In other words, it is not a time-sensitive application and can often exploit periods of relatively low-transmission activity within a network.

There are also additional demands on such multimedia systems. Besides the obvious multimedia interactivity these should be able to handle more conventional communications such as telephone calls, faxes, voicemail, and remote copying and printing of specific documents, some of which may be generated and reviewed during real-time multimedia conferences. This should not present a problem as such because the bandwidth designed for handling interactive multimedia will be more than adequate for such traffic, but the software and storage facilities to handle these transactions must be included and reserved to assure uninterrupted services.

The basic change in hardware platforms that is likely to occur is the addition of videocameras and audio/video hardware that compresses and manipulates multimedia data. The important thing to remember in this regard is to make sure that these platforms will support various operating systems including DOS, Windows, Windows NT, OS/2, and several versions of UNIX. The multimedia communications environment must handle them all without a hitch if it is to be an effective system that will be used by the employees of the company and if connectivity with outside organizations is to be established in a timely fashion when required.

Electronic Conversational Environments

Aside from the requirement for interactivity of the end user with other members of a collaborating group, there is another aspect of conversational computing that must be taken into account. It pertains to the creation and design of the electronic conversational environment at the user interface. This means manipulation of all the multimedia elements such as text, data, graphics, animation, audio, and video into an attractive and flexible interface with specific characteristics that reflect and closely simulate real conversational situations.

One of the most valuable features in this regard is the capability to display several images of participants simultaneously in windows on a single screen. Most systems limit how many such images can appear on the screen at the same time but also provide the facilities to change images in windows at will. It is also important to have a system that automatically indicates or highlights the video of the person who is actually speaking at any particular time. Clearly the extent to which a system is capable of displaying multiple images and indications about each participant will depend on the number of channels that a particular station can handle simultaneously. In the final analysis, this depends on the bandwidth of the network and the end-user connection.

This interactive conversational aspect, which includes multimedia elements, enhances current communications qualities in the communications channel and makes multiparty videoconferencing an attractive and desirable alternative.

Such electronic conversational environments also offer an additional feature that provides insights not normally available within conventional conversations. These systems can bring into one of the windows on a screen the image of end users themselves so they can monitor their own behaviors and reactions. It allows users to be fully aware of how others perceive them during such conversations.

Multiparty electronic conversational capabilities are very useful in collaborative computing in providing timely information, training, developing consensus, and arriving at specific decisions. More importantly, properly designed systems can record and store for future review and analysis all significant conversations, including the nuances of body language and voice intonations of all participating parties.

There is also some belief that such free conversational capabilities in which all members of a group can participate and leave a permanent mark on the proceedings may facilitate consensus building and decision making. These conversational features are also crucial to more private two-party or desktop-to-desktop multimedia communications between executive decision makers either preceding or following more general consensus building collaborative sessions.

Multimedia Content Issues

Real-life conversations between individuals or collaborating groups within an organization include acquisition, interpretation, and transmission of various multimedia objects. Any design of an interactive multimedia communications environment such as networks must therefore include the capability of manipulating multimedia elements such as text, graphics, animation, images, voice, music, film, and videos, as well as strictly programming-related objects such as data, code, and frames of information.

As a result, the design of an interactive multimedia communications system must include at an early stage a determination of the types, volumes, and sources of multimedia objects that will be manipulated and integrated in the networks. Such planning permits the inclusion of appropriate hardware and software tools and devices to acquire, manipulate, and transmit all the required multimedia elements in an optimal fashion and to determine the bandwidth of various network transmission segments.

Such an approach also facilitates the determination of ownership of many multimedia elements that will be included in the system. This is particularly important because if multimedia elements that are being used come from sources outside the organization, permissions and licenses may have to be obtained before such items can be incorporated in a system.

The inclusion of images, voice, and video elements also implies massive storage requirements and specialized multimedia databases, filing systems, and client-servers. These are new solutions that are specifically designed for handling massive data files such as BLOBs that are representative of many multimedia transmissions.

Networking Integration

Most networks such as LANs that currently exist have been developed to handle conventional information processing consisting primarily of text and data. Although many LANs are now being integrated into WANs, they represent arrays of different platforms, operating systems, and application areas. Many are seen as the proverbial islands of automation that are incompatible with other LANs deployed in other parts of the same organizations. This situation is aggravated only if integration of LANs in different organizations is necessary, as would be the case when a virtual corporation operation is being developed.

Interactive multimedia communications implies access and transmission of image-based information between various sources and locations. As a result, it is imperative that associated networks and transmission facilities include appropriate interfaces and devices to handle high-speed broadband multimedia traffic. Seamless integration of different networks in order to meet these objectives is one of the most challenging tasks in building these systems.

One of the most promising solutions of this problem is the *asynchronous transfer mode* (ATM) networking technology, which can provide incremental growth for existing networks as these are integrated into interactive multimedia communications systems. ATM is a unique technology that works seamlessly across the enterprise handling any multimedia content from desktops to different WANs. It is the ideal solution for public and hybrid networks, and its capabilities range from local areas to global connectivity.

User Interfaces

General-purpose interactive multimedia communications environments must be equipped with very user-friendly interfaces that can be used without effort by even the most computer-illiterate workers. This means a GUI with very simple controls that can be understood and manipulated by anybody. It must be assumed that despite computer training and experiences at the earliest age, significant segments of the working population will have little interest in computers and an even larger percentage will remain computer-illiterate.

As a result, user interfaces must be completely transparent, allowing anyone to initiate a multimedia conference within minutes by providing facilities to identify participants, specifying the type of conversation required, and outlining the subject matter for discussion. Figure 2.3 is an example of a relatively simple setup screen of this type which is, in fact, quite common in current multimedia conferencing systems. The critical aspect here is simplicity of the user interface, and it must be kept in mind that more complex interfaces will inhibit many potential users from becoming involved with such systems.

The question of the user interface is not a trivial one as it is often seen among information processing professionals. When an interactive enterprise and virtual corporation are being contemplated, the user interface must be attractive and simple to use to large, diverse, and often unpredictable user groups from outside the information processing community within and without the corporation.

The problem can be solved by developing alternative interfaces that are specialized for specific end-user populations. Such an approach may be very expensive, particularly when actual user groups of the system cannot be predicted with any certainty at the beginning of the development program. In that type of environment the best approach is most likely a common user interface that is acceptable and easily understood by the greatest number of potential end users.

On the other hand, many GUIs that already exist within various operating systems allow customization of the user interface by individual workers. Once multimedia networks are well established within an organization and usage costs come down, special groups will customize the interfaces with features and

Multiparty session

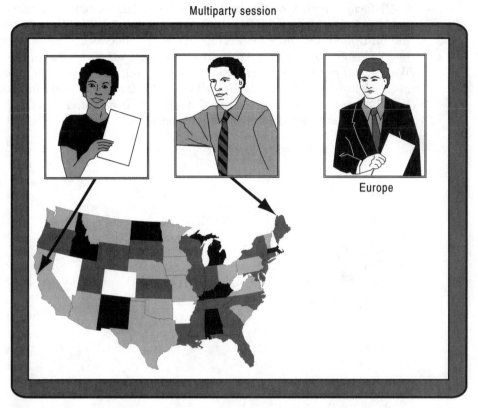

Figure 2.3 Typical collaborative screen interface. (*Source: National Semiconductor.*)

facilities that best reflect their natural ways of operation. The only caution in such instances is to make sure that the customization process does not interfere or degrade the overall collaborative objectives and that it lends itself easily to periodic updating of the system.

Real-Time Networking Requirements

All interactive multiuser conferencing systems are time-sensitive applications. This means that such transmissions must operate in real time with very low latencies or transmission delays because end users will not tolerate any noticeable delays in two-way communications for any significant length of time. In practice this also means that various interfaces, hubs, and other networking devices may have to process multimedia data faster than real time (FTRT) in order to create the perception of real-time operations for the actual end users. Fortunately, developments in digital signal processors (DSPs) and availability of faster processors at continuously decreasing prices hold promise that such systems are achievable and will be cost-effective.

All forms of multimedia conferencing are time-sensitive applications and automatically create tremendous data throughput demands on all the computing and telecommunications components of a network. The problem is even more serious when such transmissions take place across different LANs, WANs, and public networks. On the other hand, once the major elements of a network are designed to handle such real-time traffic, all other interactive multimedia applications that are not time-sensitive such as kiosks, training, and file sharing can be accommodated with ease.

Chapter

3

New Media Ventures and Businesses

Worldwide publicity given recently to the emergence of the interactive "new media" communications era is stampeding companies large and small to look for a stake or an entry point into these new markets. As a result, there is a rush of activity among long-established multinational corporate giants, small specialized vendors, and new ventures that are being formed in anticipation of capturing a meaningful market share in one or more of the rapidly growing market segments.

This activity is fueled by continuously increasing attention given to the announcements of huge development programs designed to facilitate interactive multimedia networking and communications. These include the national digital information superhighway, the 500-channel interactive TV potential, and the frantic mergers, acquisitions, and alliances taking place between major telephone, cable TV, entertainment, and information processing companies. Some business analysts compare this new environment to the previous NASA (National Aeronautics and Space Administration) space programs and major weapons development projects of DoD (U.S. Department of Defense), which only escalates the merger and acquisition frenzy among potential participants.

At stake are not only new markets with an estimated potential of several hundred billions but also leadership positions among business segments within a new world order of modern multinationals. Global interactive multimedia communications promises a cost-effective infrastructure to influence governments, minimize taxation, exploit lowest-cost production, sway public opinion, and control technologies and information to capture market shares and maintain a competitive edge.

The complexity of interactive multimedia networking and communications in business, industrial, consumer, and educational environments is reflected in the range and variety of companies involved in these machinations. These new mergers and alliances are also giving rise to the coming together of corporate entities that basically mistrust each other but are forced by developing circumstances to collaborate more closely or lose control of their existing markets or even their businesses. This is particularly true between the well-established broadcasting, publishing, and entertainment empires as well as entrepreneurial ventures of the Hollywood brand, and the technologically based telecommunications, computer hardware, and software organizations.

In fact, there are five distinct industry streams that contribute to the development of the interactive multimedia communications, each of which presents a unique set of specialized market segments. (See Chap. 8 for detailed discussion of interactive multimedia markets.) These include the basic specialized components, hardware systems, software tools, distribution networks, and a huge and widely diversified end-product and service industry (see Table 3.1). All these segments are interrelated and dependent on each other in varying degrees. It is not unlikely that leading players among the new interactive multimedia multinationals will try to capture control of critical entities in all the market segments involved and maintain dominant market shares by dictating standards, selecting end products, and choosing most lucrative target audiences.

At the same time, the nature of end products and production facilities involved lend themselves extremely well for being created and delivered through a multitude of virtual corporate operations as defined in previous chapters. This clearly means that nimble organizations that will develop such capabilities stand to gain the most market share faster than their competitors in such volatile new media market environments.

TABLE 3.1 The Multimedia Industry Structure

Industry segment	Multimedia products and services
Components	Audio/video compression microchips, ASICs, multimedia DSPs, fast microprocessors, massive memory chips, fiberoptic components
Hardware	Computers, massive storage systems, CD-ROM players, rewritable optical disks, audio/video boards, videoservers, codecs, scanners, videocameras, high-speed networking interfaces, intelligent hubs, digital switches, MCUs, TV set-top controls, cellular phones, PDAs
Software	Multimedia operating system extensions, authoring systems, multimedia databases, multimedia development tools, client-server software, ODBs, video games, movies, advertising, broadcasting, publishing, movies, multimedia content developers
Distribution	Telecom services, LANs, MANs, WANs, internetwork carriers, satellite, wireless, cable TV, CD-ROM disks
Applications	Business, education, and consumer markets

A Vision of Massive New Markets

Companies are positioning themselves as best they can to take advantage of what they perceive as massive new markets that will come about as a result of the introduction of digital interactive multimedia technologies synthesized into specific services within the information superhighway concept.

Among the most active are cable TV, telephone, and entertainment companies because many feel that their traditional markets and ways of doing business are being threatened by the new developments. The perception of 500 interactive channels on cable TV networks and all that this implies is a major incentive to them to get involved as early as possible.

Those companies know that once an interactive multimedia communications infrastructure is in place, those who control the means of distribution and content supply can look forward to an evergrowing repeat business for decades to come. They also feel that the massive consumer on-line service markets will finally become a profitable reality whose potential annual revenues are seen in the order of $120 billion in the United States alone.

Catalog shopping, which is estimated at $51 billion annually, is seen as one of the first target markets to exploit the interactive multimedia technology through existing specialized videotext consumer networks and enhanced cable TV systems for handling a variety of services on demand.

Catalog shopping firms, of which an estimated 10,000 exist, are the largest business segment considered to be a prime target for interactive multimedia technology introduction. Other markets that are equally threatened include broadcast advertising, home video rentals, information services, music records, tapes, and CDs, movie theaters, videogame arcades, cable advertising, electronic messaging, and conventional videoconferencing.

All these are consumer-related services and products, and in the earlier days of multimedia development they were thought to be the opportunities that will drive the rapid and massive introduction of interactive multimedia technologies. The realities of the marketplace changed those perceptions somewhat when it became clear that all those visions of future interactive bonanzas depend on the existence and availability of a cost-effective digital broadband transmission infrastructure which will take some time to put in place.

As a result, more attention is now being paid to development of interactive multimedia communications and networks within the corporate and institutional environments. These efforts include the introduction of enterprisewide videoconferencing, group collaborative computing networks, just-in-time training systems, permanent and portable business presentation and marketing systems, and enhanced customer services as well as various forms of concurrent or simultaneous engineering concepts in the manufacturing industries.

The massive consumer markets potential, which is increasingly being discussed under the catchall terminology of interactive TV, is nevertheless an extremely powerful incentive for development of new products to advance and facilitate the introduction of interactive multimedia networks and communications in all sectors of the economy. This is understandable because the basic technologies required for storage and transmission of multimedia traffic are identical in the commercial and consumer environments.

The separate market visions combine to create the perception of a whole new "multimedia industry" taking shape. It includes the design and manufacture of specialized multimedia processing microchips and components, assembly of multimedia-capable hardware and peripherals, development of a plethora of software tools, construction of the appropriate distribution infrastructures, and the creation of endless end-user products and services.

Among the major players in this new industry are semiconductor, computer, consumer electronics, videogame, software, broadcasting, publishing, entertainment, advertising, cable TV, wireless, telephone, and videoconferencing suppliers. The new and significant factor, however, is the fact that now most of those vendors must act together rather than operate in distinct self-contained markets. Growing interdependence between such entities is responsible in large part for much of the strategic positioning, merger, and acquisition activity between the major players in these industries.

The Microchip Manufacturers

The most critical component which makes multimedia processing a reality is the microprocessor that is fast and powerful enough to handle massive digital data representing various multimedia elements. Semiconductor manufacturers realized that they must supply increasingly powerful and specialized microchips to enhance PCs and workstations as multimedia-capable platforms.

At first vendors developed special data compression microchips for use with audio/video boards to provide multimedia capabilities for existing platforms. This approach was only a temporary stopgap measure, however, and presently such peripheral hardware devices are being designed around specialized DSPs. These are basically specialized coprocessors, and the latest versions can handle whatever multimedia processing function is required by downloading an appropriate algorithm. DSPs are discussed in much more detail in Chap. 16.

Semiconductor manufacturers perceive DSP markets as a major future business and are developing new multimedia-specific microchips in collaboration with leading computer hardware manufacturers such as AT&T, IBM, and Apple. Overall DSP sales have been increasing at over 30 percent annually, and multimedia-related DSPs are the fastest-growing segments providing devices for handling compression, audio/video codecs, and voice and music applications. Most leading semiconductor manufacturers are already involved in DSP design and manufacture, and smaller firms are entering the market with specialized multimedia devices.

Computer Hardware Vendors

Few computer hardware platforms have been designed to handle multimedia data processing, but during the late 1980s some vendors began perceiving multimedia as an enabling technology that will increase the value and usage of computing hardware. As a result, numerous add-on hardware peripherals were developed in form of audio/video boards, laser disks, CD-ROM drives, speakers, high-resolution displays, and touch screens, many of which are offered bundled in multimedia upgrade kits.

It is also necessary to distinguish between the major market segments that are emerging, in all of which computer hardware manufacturers will play a major role. The corporate end user is the established and well-defined market that is the basis of this growth and will be expanded in the future by increasing the percentage of multimedia-capable PCs and workstations in corporate networked environments. The new significant but unproven market segments with massive growth potential are the multimedia storage and distribution infrastructures and the consumer devices that will turn any TV set into an interactive multimedia terminal.

Originally the objective was to take advantage of existing large populations of PCs and workstations and provide upgrading packages which would allow end users to protect their investments in existing hardware. Only in very few instances, as in the case of Commodore Amiga, the PC platform was designed to handle multimedia data at the outset. Most hardware vendors, large and small, relied on development of upgrading kits until the early 1990s when built-in multimedia capabilities became an option available with new PCs and workstation hardware.

More recently it became obvious that the future of multimedia technology is linked up with the ability to provide interactive networked applications. Major computer hardware vendors now realize that their technologies are crucial to the development of interactive multimedia services in massive consumer markets as well as in traditional business and manufacturing environments. As a result, companies like IBM, Hewlett-Packard, and Silicon Graphics are seeking strategic alliances with major players in the semiconductor, telecommunications, and entertainment industries simultaneously.

Initial market research suggests that about 80 percent of all multimedia platforms will be IBM and IBM-compatible PCs using DOS, Windows, and OS/2 operating systems. Such platforms are expected to dominate corporate presentation, training, referencing, sales, and marketing applications, most of which will operate in enterprisewide networked environments. Another 15 percent of the total are seen as Macintosh platforms traditionally well represented in multimedia production, interactive kiosks, and desktop publishing. These percentages are expected to define the market for some time to come, but increasingly multiuser operating software, multimedia applications, and networking interfaces are being designed to handle transparently the equipment of all vendors through various open-systems schemes.

Computer hardware vendors are also moving to exploit the growing multimedia conferencing markets which are basic to the establishment of enterprisewide collaborative systems. They see an opportunity to replace expensive and specialized videoconferencing hardware systems with much more cost-effective desktop conferencing products and services. At the same time, manufacturers of powerful workstations see new markets and opportunities for their products as client-servers for multimedia LANs, concurrent engineering, and merchandising kiosks in networked multimedia applications.

Link Resources, a market research firm, estimated multimedia hardware market at $2.8 billion in 1993 growing at an annual rate of nearly 70 percent and expected to reach $8.4 billion in 1996. With increasing interest in interactive multimedia communications sparked by the promotion of the information superhighway concepts, these markets may grow even faster than expected.

Software Developers See a Broader Market

Once multimedia communications comes into play, software takes on a meaning much broader than that in conventional information processing. On one hand, operating systems, multimedia development tools, and networking software must be enhanced with multimedia processing capabilities. On the other hand, multimedia content in form of documents, images, films, videos, books, artwork, music, and voice must be digitized, compressed, stored, indexed, and manipulated. What this means is that in addition to computer software vendors, there are publishers, TV, movie, broadcasting, videotape, and archival companies that have a stake in the interactive multimedia communications markets.

As far as basic operating systems are concerned, software manufacturers are competing to acquire the largest possible market share with multimedia capable operating systems that can handle the interactivity with most installations in an enterprise. The major operating systems involved are DOS, OS/2, Windows, Windows NT, and Netware, all of which are being developed with various multimedia processing and networking capabilities. In addition, there are numerous UNIX versions with significant numbers of networking users that are important potential markets for interactive multimedia communications. These include SunSoft Solaris, SCO UNIX, HP-UX, IBM AIX (Advanced Interactive eXecutive), UNIXWare, NextStep, and many others with over 500,000 additional installations. The most popular operating systems are expected to be OS/2 and Windows NT with 7,500,000 and 5,000,000 units forecasted in operation by 1997. As a result, IBM, Microsoft, and Novell—which now controls UNIX Systems Laboratories—are the three leading contenders for supremacy in this market segment.

Multimedia software tools include authoring systems, graphics, audio and video editing, animation, multimedia file managers and databases, specialized client-servers, and a whole gamut of multimedia design utilities. These are used for digitization and management of multimedia context, screen capture, image enhancement, and special effects. Some of these products have been developed by major hardware manufacturers, notably IBM, but the trend is now for new specialized multimedia software ventures to develop such tools and team up as business partners with major hardware vendors.

Another almost unlimited market opportunity is seen by multimedia application developers who combine specific industry or functional expertise with the most appropriate development tools to produce final solutions for users. They supply multimedia business presentations, training courseware, and specific sales and marketing applications for deployment in corporate and public networked environments. Increasingly, because of the multimedia content requirements, there are new ventures entering this market combining computing technology with creative multimedia capabilities.

Traditionally software vendors are also positioning themselves to influence multimedia processing and transmission standards. A major role in this respect has been played by Microsoft, the largest software company in the world. It is involved with Tele-Communications and Time Warner in Cablesoft, a joint venture designed to establish a standard for the next generation of interactive multimedia services. It is also teamed up with Intel and General Instrument in design or critical multimedia software for TV controllers to manage interactive TV services for the massive consumer markets.

The Future of Telecommunications Is at Stake

The most active corporate players positioning themselves to control interactive multimedia distribution networks are telephone and cable TV companies. Both types of organizations provide telecommunication services and play a major role in financing and constructing various digital transmission facilities that form the elements of the information superhighway of the future.

The telephone companies control access to all business, government, institutions, and practically 100 percent of U.S. households, although many of their facilities do not have broadband transmission capabilities. Cable TV companies, on the other hand, already provide services to over 60 percent of all households with a significant percentage of their distribution networks with broadband fiberoptic transmission capabilities. Cellular telephone and wireless communications services are also interested players in the potential of interactive multimedia communications.

It is expected that by the year 2000 about 40,000,000 households will be linked directly by fiberoptic networks to the information superhighway. Companies who will control access to those markets will be major players in the interactive multimedia services business.

The telephone companies have an advantage because they are ahead in digital switching technologies and generally control more significant financial resources than do the cable TV services. On the other hand, telephone networks connect most households with low-bandwidth copper wires which are unsuitable for efficient transmission of interactive multimedia services. As a result, telephone companies are expanding their broadband fiberoptic links, developing compression schemes, and acquiring or teaming up with cable TV companies to exploit their existing broadband transmission facilities. Figure 3.1 suggests how the households of the future may be connected to the information superhighway.

The $25 billion merger of Bell Atlantic and Tele-Communications announced in 1993 is a prime example of this activity and underscores the magnitude of future business that is at stake. Although that particular merger did not go through because of extensive cultural differences and pricing market issues, it is an indication of the visions entertained by many executives and industry analysts. Actually over 200 various mergers, acquisitions, and joint ventures between service organizations of this type have been reported in North America in recent years. Table 3.2 summarizes some of the more prominent deals that have taken place or are under way.

Multimedia Content Control

Broadcasting and motion picture production companies are by their very nature already in the multimedia business, although the traditional mode of delivery does not include interactivity. Companies like ABC, Columbia, Paramount, or Time Warner control large archives of intellectual properties such as movies,

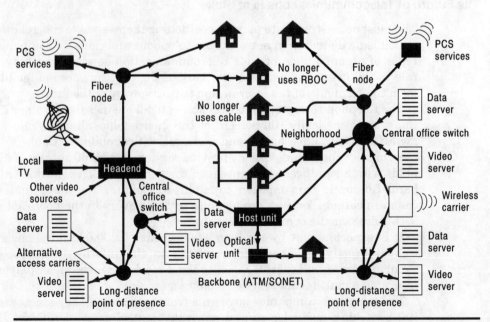

Figure 3.1 Information superhighway connectivity. (*Source:* BYTE, *March 1994, p. 49, copyright by McGraw-Hill, Inc.*)

TABLE 3.2   Multimedia Activities of Telecommunication Companies

Telecom company	Details of interactive multimedia activities
Ameritech	Rebuilding phone networks into interactive video pipelines using DEC videoserver technology
AT&T	OneTouch video-on-demand pilots with Viacom in Littleton, CO, and Castro Valley, CA Imagination Network is entry into interactive multimedia entertainment through investment in The Sierra Network Acquired McCaw Communications National Lotus Notes public server network Switching systems for Full Service Network interactive TV pilot in Orlando, FL
Bell Atlantic	$33 billion supermerger with Tele-C-Communications canceled Testing video-on-demand services with BT of UK Interactive TV pilot in Virginia uses Oracle videoserver technology
Bell South	Invested $250 million in Prime Cable and has equity in QVC network for home shopping
BCE Telecom International (Canada)	Acquired 30% of Jones Intercable the 8th largest cable TV operator for $400 million
NYNEX	Testing video-on-demand services in Portland (ME), New York, and eastern Massachusetts; offers MBS multimedia transmission services
MCI Communications	Developed international videoconferencing services; BT owns 20% of company
Pacific Bell	Developing high-capacity interactive network for multimedia services in California by 1996 using Hewlett-Packard videoserver technology
Southwestern Bell	Acquired Hauser Communications cable TV systems for $650 million in Washington, DC, area and invested in 21 cable TV systems of Cox Enterprises
US West	Tested video-on-demand in Omaha, NE using DEC videoserver technology; holds 25% of Time Warner Entertainment and is upgrading cable TV system for interactivity; talks joint ventures with Cablevision Systems

documentaries, photographs, music, and various forms of databanks. They are also best equipped to create and manipulate multimedia content, which is a critical aspect of interactive multimedia communications, with which few conventional information processing firms and personnel are familiar.

Current trends in the entertainment industry also suggest that there is a shift in audience viewing customs from predominantly advertiser-supported productions to consumer-paid and pay-per-view delivery modes. Broadcasting organizations are therefore naturally attracted to interactive multimedia networks because these are relatively inexpensive methods to provide entertainment on demand to all types of audiences.

In practice this means a faster and greater return on their initial investments in programming. This is an important factor because as time progresses, the sunk-cost of entertainment products is increasing. Interactive multimedia networks also provide a means to measure audience demographics and preferences with greater precision than hitherto possible. This in itself is a factor that contributes to faster and greater return on their investments.

Some of the interactive TV system developers such as Eon, Inc. (formerly known as TV Answer) and Interactive Network have made it clear that this approach can provide advertisers with precise information indicating which particular seconds of a commercial were responsible for generating a purchase order or specific product interest.

All these are new and extremely powerful reasons for broadcasting organizations to move vigorously into the world of interactivity. Deregulation, which allows TV networks and movie studios to acquire, own, and operate cable TV and other distribution services, is also a major factor in motivating those companies to take action.

Publishing Companies

Besides broadcasting and movie studios, numerous publishers of books, magazines, newspapers, and databases also control large quantities of information. These sources are of particular value for design and development of integrated multimedia training and educational applications as well as in marketing, investment, and business development. Such information is already being delivered in electronic form in many cases in form of CD-ROMs on standalone platforms or via interactive networks on just-in-time basis.

Initial steps have already been taken by many publishers who developed CD-ROM-based versions of their popular encyclopedias and other specialized publications. In such cases images, video, sound, or music are of importance in transmitting the knowledge or providing effective training.

Several thousand CD-ROM titles have already been published estimated by Infotech at over 4600 in 1993, and production of such products is growing at a rapid rate. Frost & Sullivan estimates that reference databases represent 38 percent of those, games 29 percent, and edutainment (educational entertainment or documentaries) 25 percent. Most of those CD-ROM databases include directories, lists, and statistics developed by organizations such as McGraw-Hill's DATAPRO, Dun & Bradstreet Marketing Services, TRW Business Credit Data, and U.S. Geological Survey. Reuter, which provides various sets of financial statistics worldwide in electronic form, is already planning to include multimedia components with its services. Networked CD-ROM databases have been set up that offer remote access and search facilities including usage statistics monitoring capabilities that are of great value to service providers.

Simultaneously there is a growing number of CD-ROM platforms either within PCs or in specific players such as CD-I, CDTV, or 3DO products that use standard TV sets as monitors. Publishers of all categories are well aware that once interactive networks become more common, their databases will have considerably larger markets to exploit. As a result, they are enhancing their existing text and data offerings with multimedia content and designing new products in form of electronic books and magazines. They are also exploring the possibility of on-line distribution of the CD-ROM-based information directly to conventional bookstores, video, and music stores who are interested in interactive multimedia promotions, presentations and distributions.

Network-Specific Products Suppliers

Many relatively small firms specialize in network-specific hardware, software, and services. They supply LAN interfaces, gateways, bridges, routers, switches, intelligent hubs, multipoint control units (MCUs), and other devices. They also provide networking software products, including operating systems, network design, testing, development, and management tools as well as client-servers and databases.

This group of vendors includes leading network products suppliers such as Novell, and Banyan; database vendors like Oracle and Sybase; and numerous smaller firms. All these traditional networking products suppliers face the immediate issue of upgrading their existing hardware and software products to handle multimedia traffic on the corporate networks. Since the original designs have not been developed to handle multimedia data sets and broadband transmissions, this process may turn out to be costly and difficult to accomplish without extensive redesign of their products. There is also a new multimedia networking products market that is associated with handling massive videoserver systems and interactive selection and billing in the forthcoming interactive TV segments on the information superhighway.

This also creates new opportunities for ventures such as Gain Technology, Starlight Networks, or Videoserver to enter the market with brand-new networking hardware and software solutions specifically designed to handle multimedia traffic. These newcomers are not burdened with prior investment and client bases that must be supported and can devote all their attention and resources to the design of competitive multimedia-capable products. As a result, the established suppliers are very vulnerable to such competition and there is considerable pressure on them to come up with better and more cost-effective multimedia capable solutions by enhancing the existing networking infrastructures.

This is not a trivial task, because multimedia networks must handle several magnitudes times more data often much faster and in real-time environments. The established vendors, therefore, face the problem of coming up with upgrades to their existing solutions that can meet such requirements at a competitive cost and offer expansion potential for the future. That is the only way they can assure their clients that their investment is protected for a reasonable period of time.

These market realities extend to all the vendors including client-server and database suppliers who must enhance their products to handle the so-called binary large objects (BLOBs) which represent massive and unpredictable multimedia traffic. As a result, there is intense competition among the vendors to develop multimedia-capable products as soon as possible and at a reasonable price. The alternative is a rapid loss of market share and possibly incursion of operating losses leading to failure.

Existing suppliers are taking different approaches to solve these problems depending on their technological and financial resources. Sybase, for example, upgraded its relational database and client server product lines by acquiring Gain Technology in 1993, which automatically gave the company a networked multimedia ODB capability with a set of development tools to go along. Oracle, by contrast, is redesigning its database products by stages providing multimedia capabilities and recently announced its Media Server as an extension of the Oracle 7 database product line.

Nevertheless, there is also a lot of activity in the venture capital community researching and financing new ventures with promising multimedia-capable products that can compete successfully in the new marketplace. This area represents currently one of the most fertile venture capital investment opportunities.

The Consumer Electronics Frenzy

Some of the most frantic activity with regard to interactive multimedia communications is taking place in the consumer electronics arena. This market includes manufacturers of TV, radio, videogames, hi-fi (high-fidelity) audio, CD-ROM players, home computers, conventional and cellular telephones, and more recently personal digital assistants.

Many of the vendors are under the impression that interactive multimedia consumer markets will turn out to be the largest and most profitable of all. Others, like IBM, have concluded after an initial investigation of the issues that the business and institutional environments present more immediate opportunities for design and installation of interactive multimedia communications systems and services in the short run.

Nevertheless, the consumer multimedia business potential sustained by publicity surrounding the information superhighways is an important factor in promoting and accelerating developments in the corporate world. Key players include computer companies, communications services, household equipment suppliers, and on-line service vendors like Prodigy. Most expect that sooner or later they will be in a position to offer new products to the consumer which may include interactive TV sets, multimedia home computers, CD-ROM players, videogames, or various interactive services in the edutainment category. There is little question, for example, that leading videogame competitors like Sega and Nintendo are bringing to market increasingly more sophisticated interactive multimedia networked products and companies like Matsushita, Philips, and Sony are also actively developing such new products for the consumer.

There is also a group of new ventures that are experimenting with new interactive products and services targeted at the consumer which will require sophisticated multimedia networking as a backbone infrastructure. These include such concepts as interactive TV, interactive movies, infotainment, multiparty videogames, videotelephony, electronic video stores, and various virtual-reality marketing services.

The interactive consumer electronics market is in turmoil at present because the basic issue of "how much interactivity will couch potatoes accept and require" has not yet been resolved. Another retarding factor is the lack of standards because consumers tend to prefer electronic gadgets that can perform with a variety of existing input media. This concern is equivalent to the protection of prior investment in the business environments.

An associated issue that is very critical in consumer electronics is the price of products and services, which must be within an affordable range for the largest possible audiences without disproportionate operating burden on the suppliers. The profitability of many on-line consumer services so far remains questionable, but many industry observers expect that the inclusion of multimedia content and interactivity into such services will greatly increase their appeal to the consumer.

Some market research organizations, such as Link Resources of New York, are skeptical about the size of the multimedia consumer markets and estimated the total at about $1 billion in 1993. According to their forecasts, however, significant growth at average rates above 100 percent per year can be expected in the immediate future, which would bring those markets to the $6 billion level by 1996. This would suggest that by the mid-1990s multimedia consumer and business markets would be comparable in size. The massive consumer electronics market potential of the future provides a powerful incentive for development of cost-effective interactive multimedia communications products and services. Nevertheless, most industry analysts believe that corporate and enterprise applications such as desktop conferencing, sales presentations, and concurrent engineering are the main driving force in development of the information superhighway businesses.

A Bonanza for Systems Integrators

Design of interactive multimedia communications systems, particularly with real-time collaborative groupware components, presents complex and specialized challenges. A variety of new skills and leading-edge technology know-how required to conceive and implement such systems is not always available within a corporate environment. As a result, companies often rely on third-party consultants and system integrators who can analyze existing networking infrastructures, suggest the most appropriate solutions, and implement the system for their clients.

Two categories of consulting organizations will benefit in these new markets. First there is a group of high-level management consulting firms who are specialists in evaluating existing corporate organizations and can suggest what changes must be made to maintain market share and competitiveness as a result of introduction of more advanced technologies such as interactive multimedia networking and communications. Those are the consulting firms that specialize in corporate change. They will conceive and design the interactive enterprise strategies and map out the evolution into a virtual corporate operating environment.

Some of those consulting organizations are the management consulting divisions of the "big six" accounting firms, such as Andersen Consulting, Coopers & Lybrand, or Deloitte & Touche. Others are independent management consulting firms such as Booz Allen & Hamilton, McKinsey & Company, or The Boston Group.

Another group of consulting firms bases their activities on the rapidly changing information technologies at the operating level. These companies provide more practical solutions by supplying teams of specialists who can design and implement interactive multimedia networks and communications systems on a turnkey basis. These companies often work in conjunction with management consulting firms and include such organizations as American Management Systems, CAP Gemini Sogeti, Electronic Data Systems, SHL Systemhouse, and many others, including consulting divisions and subsidiaries of major computer hardware and software firms.

System integrators of this type are keen promoters of interactive multimedia communications technologies and a major force in implementing change and innovation. During the last few years these companies have been increasing the availability of interactive multimedia skills within their organizations. Considering the complexity of multimedia networking, it can be expected that most systems integration firms will be in this business before long and new, more specialized niche consultants will also appear in the market.

4

Multimedia Conferencing

The multimedia conferencing business is taking shape as a result of the coming together of new advanced videoconferencing hardware, adoption of a number of pertinent video compression and transmission standards, and massive transition toward specialized networks and client-server technologies. Multimedia conferencing offers a strategic advantage to corporations that can exploit the technology with creative group automation initiatives on a project, corporatewide and global basis.

The trend to automate corporate group functions is leading to the integration of videoconferencing and networking technologies into more effective multimedia conferencing systems. Rapid progress in basic technologies makes it possible to expand conventional point-to-point videoconferencing into multipoint multimedia systems that offer strategic advantages to the corporate users.

Several facilitating technologies are advancing the cause of multimedia conferencing and are bringing it to the desktop level. These include commercial development of programmable video-specific DSP chips, microminiaturization of electronic cameras, adoption of more efficient compression standards, better codec hardware, and development of specialized multipoint control units.

About two dozen vendors have positioned themselves to take advantage of the emerging multimedia conferencing markets. These include specialist videoconferencing hardware ventures, major telecommunications carriers, leading semiconductor firms, and several large computer manufacturing companies.

A Major Multimedia Application Objective

Multimedia conferencing is a major aspect of collaborative systems and is sometimes seen as the "killer" application of multimedia technology. Such collaborative multimedia systems are seen as the most powerful means of communications for the interactive enterprise of the future. As the new corporate paradigm for the 1990s, it is expected to eliminate the barriers of time and distance for more effective collaboration between corporate employees and their clients.

The vision of the interactive enterprise as it is seen by its proponents ultimately includes collaboration of any corporate worker with anyone, anywhere in the world regardless of time, geography, computer platforms, and applications. Through the use of multimedia elements in such real-time transmissions it is possible to conceive of a "virtual conference room" where marketing, R&D (research and development), outside suppliers, distributors, and even customers can collaborate on a project without leaving their offices in various parts of the world.

It is now becoming quite clear that the future of multimedia technology depends on the development of group multimedia systems which provide better communications and more competitive ways of working together. These concepts include real-time interactive collaboration sharing screens and whiteboards, with personal videoconferencing and multiple-way communications.

Multimedia conferencing within those parameters can take several forms which basically fall into the point-to-point and multipoint categories. Within the point-to-point category multimedia conferencing can involve two persons conferencing interactively through their PCs or groups of persons communicating from one multimedia conference room or facility to another. Multipoint multimedia conferencing involves three or more locations which can be single PCs, PC LANs, or multimedia conference rooms and is also known as *group conferencing* within a project leading to more extensive corporate and global conferencing systems (Table 4.1).

The challenge of multimedia conferencing is to provide real-time interactive connectivity at all levels. These include individual workers, collaborating groups within a single LAN, and enterprisewide collaboration between various LANs within and without the corporation. The necessity to collaborate with individuals and groups outside the corporate environment presents the biggest challenge, but it also provides the real means to operate as a virtual corporation.

Point-to-Point Videoconferencing

Significance of standalone multimedia platforms

Currently the most common use of multimedia technology is in the form of standalone multimedia platforms. These systems provide more effective ways of delivering information and knowledge for individual workers and an increasing number of such platforms are being connected within LANs and even WANs, although without videoconferencing capabilities. As time progresses, such multimedia-capable networked platforms are prime candidates for introduction of multimedia conferencing. The major inhibiting factors so far are lack of sufficient bandwidth on many LANs, and relative scarcity of multiprotocol products such as client-servers, intelligent hubs, bridges, gateways, and

TABLE 4.1 **Multimedia Conferencing System Categories**

Type of system	Conferencing facilities	Description
Point-to-point conferencing	Videophone	Consumer-level two-way video communications over standard telephone lines with relatively poor quality and small-size video of 2–10 fps
	Desktop-to-desktop	Involves PCs in two separate locations; videocameras may be optional with voice teleconferencing and shared screens of data, text, or images
	Rollabout conferencing	Consists of a movable videoconferencing system that connects through modems with other rollabout or permanent videoconferencing facilities
	Boardroom conferencing	Groups gather in special conference rooms equipped with videoconferencing transmission facilities and requiring higher-bandwidth transmission lines
Multipoint conferencing	MCU-based videoconferencing	Three or more locations can be connected through the use of MCUs that can handle various codecs between different conference facilities
	LAN-based videoconferencing	Three or more PCs or workstations involved in group collaborative sessions linked through LANs and WANs and sharing screens and databases
	Concurrent engineering	Special case of LAN-based group conferencing designed to handle engineering design and manufacturing projects

routers capable of handling massive video and audio data transmissions on an interactive basis.

Videotelephony

Videotelephony is basically two-way video communication at the consumer level using existing telephone equipment and networks. It is strictly a point-to-point or person-to-person system which uses a relatively small screen and very low image frame rates, in the range of 2 to 10 frames per second (fps). As such these systems are very primitive and unsuited for use in most corporate environments, but their existence and use by the general public is seen as an activity promoting videoconferencing in all its forms throughout the economy.

Videotelephony can be considered the prototype of videoconferencing and was first introduced by AT&T in 1964 at the New York World's Fair when the

PicturePhone product was demonstrated. While the public was excited by the possibility of face-to-face conversations across long distances, the costs of these products and implementation difficulties were such that videotelephony did not become a viable product or service for many years. The first such personal videotelephone was introduced by AT&T in 1992 as the VideoPhone 2500. Other attempts at developing videotelephony were made by Japanese firms in mid-1980s using black-and-white (B&W) still images, but these products were not successful.

The videophone of today is basically a telephone with a small videocamera and screen driven by compression microchip codecs, but it does not have the general processing capability of a PC. It plugs into a standard modular telephone jack and uses a 3.3-in color LCD (liquid crystal display) video screen which can display video images at relatively low rates of 2 to 10 fps. This makes the images jerky and uneven, and the limited bandwidth of the telephone lines makes it difficult to enhance this performance, although more powerful compression microchips may improve this situation in the future.

The typical videotelephone has a special video transmission control button which switches video transmission on or off as desired. It can be used with the conventional telephone handset or speakers if group communications is required. The videocamera can focus to up to 9 ft from the console to accommodate several persons when necessary.

Whatever the limitations, videotelephone manufacturers are still optimistic about its future. They feel the performance of such devices will improve drastically once the information superhighway is in place or interactive networks become more widely available directly to the home.

Desktop-to-desktop conferencing

This form of multimedia conferencing involves only two users or two small groups of users that can cluster around a single PC platform that communicates over existing networks or transmission facilities with another PC suitably equipped to receive such communications. This concept combines videotelephony, teleconferencing, and PCs or workstations with multiuser operating systems. Video, voice, and data are transmitted and displayed in specific windows as if the participants were sitting face to face in a conference room or working side by side at a drawing board.

Such systems may or may not include videocameras to convey real-time images of the participants and often rely primarily on voice interaction for continuity. Because of the inflexion and tones communicated to the other parties on a real-time basis, such communication is considered superior to voicemail or annotated E-mail transmissions.

In desktop-to-desktop conferencing interactive multimedia sessions are used to share displays of drawings or plans and text or images of documents which are actually seen by all participants simultaneously. These systems usually involve the use of remote pointers which allow each party to point out visual details for clarification and reference. Chalkboard or whiteboard features provide capabilities for displaying simple drawings made with basic tools such as a mouse, tablet, or touchscreen supported by the system in question. Depending on input devices, images may be scanned from documents, bitmap files imported from databases, and independent notes made during the sessions. IBM's Person-to-Person (P2P) offering within its Ultimedia product line is a typical example of currently available PC-to-PC multimedia conferencing system of that type (Fig. 4.1).

Conference facility-to-conference facility conferencing

This is a more sophisticated point-to-point multimedia conferencing concept which has its roots in the original videoconferencing ideas and systems. In this type of multimedia conferencing, groups of persons gather in a specific conference room equipped to operate as a videoconference transmission and reception center. A variant of this concept consists of a movable videoconferencing system that is being wheeled to conventional conference rooms as required where it is connected through modems into the transmission networks and can communicate interactively with another permanent or mobile facility of similar type.

This type of multimedia conferencing facility depends on video compression-decompression devices (codecs) that allow two-way, full-motion, color videoconferencing at selectable bandwidths ranging from 56 kilobits per second (Kbps)

IBM's P2P software provides a core set of functions, such as a shared chalkboard, remote pointer, file transfer and optionally, a video window. These functions will make it easier to develop networked multimedia applications.

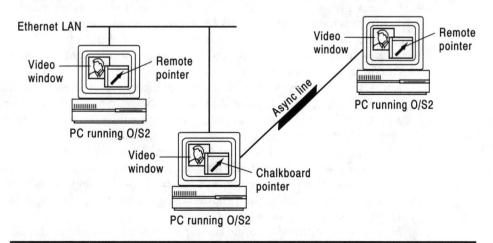

Figure 4.1 Desktop multimedia conferencing. (*Source: Reprinted from* Data Communications, *January 21, 1993, p. 27, copyright by McGraw-Hill, Inc., all rights reserved.*)

to 2.048 Megabits per second (Mbps) and enable the users to transmit video, audio, data, and graphics over a single digital channel. A codec is required at each conference facility to perform both compression and decompression functions at the site. Coded signals can be transmitted between sites over terrestrial, satellite, microwave, or cable networks. These systems may also employ extensive diagnostic features which identify system failures down to a printed-circuit-board (PCB) level and can be used from any site within the network. As a security feature, such videoconferencing systems may also offer encryption facilities with user-selectable keys that can be made unique for each transmission.

Several vendors offer integrated multimedia conferencing facilities in conjunction with their specific codec products. Such facilities usually include single- or double-monitor units, videocameras, sophisticated audio systems, and control units as well as a range of options to enable users to tailor the systems to their specific meeting requirements. Prices for such facilities may range up to $40,000 depending on the options employed.

Multipoint Conferencing

In a multipoint conferencing system participants in three or more locations can interact with each location and are able to see and hear the location which is speaking at any particular moment. This control may be voice-activated or through specific commands, but a multimedia multipoint conferencing system must additionally include the ability to transmit from any originating location various documents, images, drawings, and video clips other than those of the speakers. In addition, these transmissions should be such that they can be acted on, annotated, and changed during the presentation by any location involved in the conference.

In these systems still images may be transmitted at higher resolutions than video to provide sharper images for more detailed viewing. In addition, data ports must provide the facilities for transmission of data from devices such as PCs databases, fax machines, and camcorders to all locations on demand at high transmission speeds.

Aside from compression technology required for efficient transmission of multimedia conferencing video and data between two points, there must be a means to go beyond the two site configurations which have been the norm until a couple of years ago. Strictly speaking, multipoint multimedia conferencing can take place with point-to-point equipment controlled by human operators at a central site who can manually switch videoconferencing traffic between various locations. However, such an approach, while feasible, is certainly neither very efficient nor practical for productivity-enhancing applications such as concurrent or simultaneous engineering.

Multipoint multimedia conferencing is possible because of the development of a *multipoint control unit* (MCU), which acts as an electronic middleperson switching correct audio and video signals to all participants. These MCU products can be placed at any point in a videoconferencing network and will accept all digitized signals from codecs and automatically route them to the proper sites. The only problem so far with this approach was the fact that all units are proprietary and operate only with codecs of a particular vendor.

Startup Videoserver, Inc. of Lexington, Massachusetts, recently broke this bottleneck by developing a 28-port Model 2000 MCU which is compatible with codecs from a range of vendors and meets the multipoint videoconferencing standard H.243 of CCITT. It is the first multivendor-capable MCU product, but videoconferencing equipment vendors are also developing their MCU equipment to accept signals from codecs other than their own proprietary units.

In order to perform the multisite switching reliably in real time, MCUs are fairly complex and precise technology with powerful processors and extensive microprocessor-based software systems. All are proprietary devices typically handling 8 to 16 locations at prices ranging from $10,000 to $25,000 per port. MCUs can be bought or leased from service companies such as AT&T, Sprint, and MCI Communications, which usually handle all the proprietary codecs in use and provide the bridging facilities between videoconferencing systems of various vendors (Fig. 4.2).

Geography plays a part in MCU purchase or lease decision making. If multimedia conferencing sites are clustered relatively close to each other, it may pay to buy MCUs and set up a private videoconferencing network. On the other hand, if users are scattered widely or are mobile, use of carrier services may prove more economical. Ownership of MCUs also ensures complete control over scheduling of multimedia conference events and guarantees that the most efficient bandwidth is employed based on predominant user requirements.

Group Videoconferencing Growth

Videoconferencing and information systems developed as separate and disjointed markets since 1979. With multimedia networking concepts becoming a reality, videoconferencing users are demanding interconnectivity of their equipment with computers and databases, and simultaneous capability of transmitting and manipulating data and video elements during conferences.

Now multimedia conferencing is expected to become a requirement of international companies who want to remain competitive in the global marketplace. Desert Storm, which severely limited business travel for many corporations, turned out to be a boon to videoconferencing and proved that you do not have to travel to do business. As a result, videoconferencing is no longer seen as merely a cost-cutting device to reduce travel but as a strategic system enhancing business competitiveness.

Simultaneously prices of videoconferencing equipment and systems averaging over $100,000 in 1985 have already come down to $50,000 levels, and associated transmission costs have dropped along similar lines as a result of more sophisticated compression and transmission technologies and development of operating standards.

The BPUs inside the Model 2000 MCU handle the video switching. The network interface cards accept incoming video signals from the codecs.

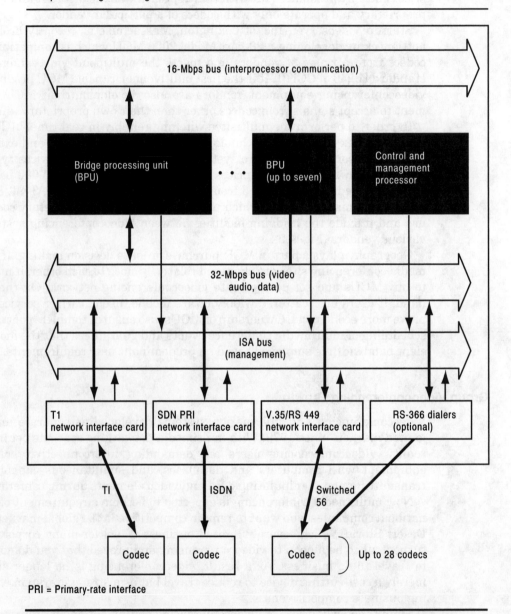

Figure 4.2 Anatomy of a multipoint control unit. (*Source: Reprinted from* Data Communications, *August 1992, p.73, copyright by McGraw-Hill, Inc., all rights reserved.*)

Since 1989 videoconferencing business has grown at 50 to 100 percent per year, and Personal Technology Research estimated worldwide videoconferencing sales at $806 million in 1993, expanding to $8.3 billion in 1997. There were less than 33,000 videoconferencing units installed worldwide in 1991, but Frost & Sullivan estimates that this can increase to over 620,000 units by 1996. This is not inconceivable, given rapidly falling equipment prices and development of personal desktop videoconferencing systems bringing economical access to large numbers of executive workers.

Reduction of travel expenses was traditionally the major incentive for videoconferencing systems purchases. Now other benefits are being discovered which suggest that videoconferencing increases information retention among participants by 50 percent and accelerates buying decisions by 77 percent. These results are based on previous psychological studies demonstrating that in face-to-face meetings only 7 percent of the meaning is conveyed by spoken words. In addition, 38 percent of the meaning is communicated by intonation and 55 percent through visual cues. These factors can be easily optimized in a multimedia conference environment and are more predictable and controllable than actual meetings.

Multimedia conferencing also appears to offer additional cost-cutting and productivity benefits in such corporate functions as training, telecommuting, job interviewing, medical consulting, distance learning, and customer services.

Concurrent Engineering

Competitive pressures in the marketplace, continual need for better quality, and shortening life cycles even as products continue to increase in complexity forced many manufacturers to seek new ways of making the product development cycle more efficient. Taking advantage of computer-aided-engineering (CAE) tools, powerful workstations, multiuser operating systems like UNIX, and communications in form of LANs, manufacturers began to attack all aspects of product development simultaneously with teams of engineers and specialists from all departments concerned. The result was the emergence of a new enterprisewide interactive methodology now most commonly known as *concurrent engineering*.

Essentially, concurrent or simultaneous engineering is group multimedia activity pertaining to specific project or product which combines desktop teleconferencing with multimedia file sharing. The need to share files became apparent during the early days of computer-assisted design and manufacturing (CAD/CAM) usage and UNIX workstations. The various stages of a design and manufacturing process were conducted by different groups of people, each generating its own databases of information pertaining to the project which had to be transmitted for use by the other groups, creating significant potential for errors and consuming a lot of time. Figure 4.3 illustrates the conventional product design process which presents costly delays when changes occur,

Traditional engineering is a step-by-step process with
each function being performed in isolation with a
permanent division between design and manufacturing.

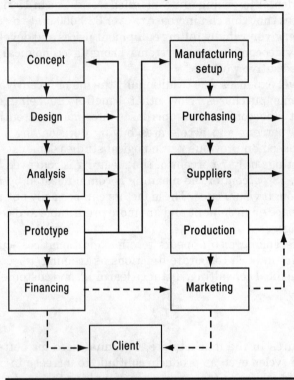

Figure 4.3 Conventional product design process.

particularly between the engineering and manufacturing functions of the process.

Integration of these files into a common database became imperative as more complex CAD/CAM systems, 3-D (three-dimensional) modelers, and rendering programs came into use, generating voluminous amounts of data. Some vendors of CAD/CAM systems also include videoconferencing capabilities in their products. The need to handle very large data objects and images led to the development of object databases (ODBs) for these applications. The obvious next step was to exploit powerful UNIX workstations already operating within LANs for coordinating the product design process with all parties involved, including sales and marketing, financing, the client, design engineers, stress analysts, manufacturing and test specialists, customer service, and maintenance. Group multimedia technology offers the means to improve corporate productivity through concurrent engineering, which is now seen as one of the most promising applications of multiuser multimedia technology. Figure 4.4 outlines the essence of this approach.

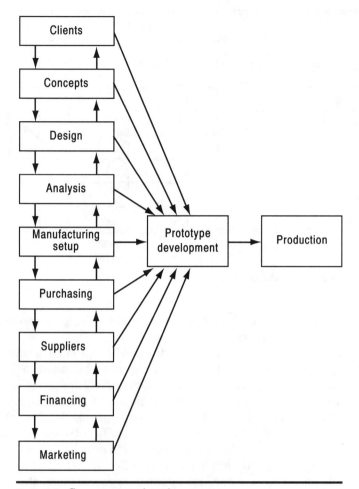

Figure 4.4   Concurrent engineering concept.

Traditional product design follows a linear path from conceptualization, through analysis, prototyping, identification of suppliers, and manufacture. Success was considered creation of a product that meets most of stated constraints and performance parameters. In such a process each step begins only after the last one is completed, and changes require costly and time-consuming redesign. Moreover, a proverbial brick wall often exists between design and manufacturing which also delays the process (Fig. 4.3)

In current intensively competitive environments there are many more constraints and standards that a product must meet in order to succeed in the marketplace. Aside from the traditional performance requirements, a product must be competitively priced, be available for purchase ahead of competition, look attractive, be easy to use and manufacture, provide customer service, have acceptable life-cycle costs, and also meet safety, health, and social standards or

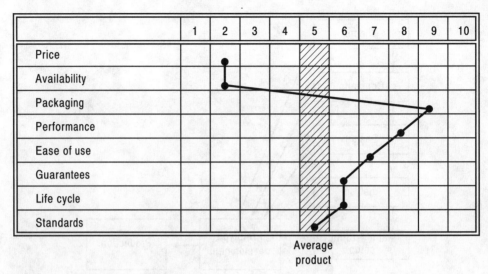

Figure 4.5   Critical product design dimensions.

perceptions. Figure 4.5 illustrates the major product design dimensions that must be brought into the process from the outset.

The concurrent engineering approach deals with all such product design aspects simultaneously. As a result, most changes can be taken care of during the early stages of the process when they are relatively inexpensive to make. This leads to fewer prototypes, shorter development times, higher product quality, faster time to market, and lower costs due primarily to elimination of extensive changes and multiple prototypes. Tradeoffs between all factors are, of course, constantly being made, but an interactive multimedia user interface within a concurrent engineering project makes the effect of any change instantly visible and permits a decision by the group to be made at once. Figure 4.6 shows the critical factors and their relative impact on overall product design in a concurrent engineering environment.

The implementation of a concurrent engineering process usually requires massive changes in a company's culture affecting all employees as well as relations with customers, clients, partners, and suppliers. Nevertheless, the benefits that are being documented and the competitive pressures of the marketplace make it mandatory for many manufacturers to take this approach. Interactive multimedia networking in these environments is in fact a facilitating technology that will accelerate the introduction and usage of concurrent engineering methodologies.

Store-and-Forward Systems

Two basic problems are associated with and limit the effectiveness of interactive collaborative applications such as multiuser videoconferencing and con-

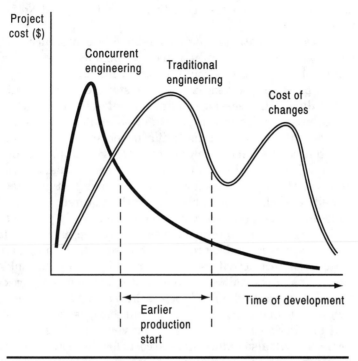

Project
cost ($)

Concurrent
engineering

Traditional
engineering

Cost of
changes

Earlier
production
start

Time of development

Figure 4.6 Impact of concurrent engineering on product design.
(*Source: 21st Century Research*)

current engineering. One is the difficulty of securing the participation of more than a few persons at the same time and is inherently a scheduling issue. The other problem relates to time differences between individuals and groups working in different time zones, countries, or continents.

The question of probability that all people required at a particular conference will in fact be able to participate is not yet fully understood. Studies indicate that the probability of any individual being able to participate within the next 7 days is about 80 percent. As a result if more than 8 people are involved the chances of getting them together at a particular hour are relatively low. With 10 people the chances drop to 11 percent, and with 25 people the probability is less than 0.5 percent. What this means is that success of collaborative group conferencing depends heavily on efficient calendaring probably using E-mail and other means of communications well ahead of time.

In the case of time-zone differences, individual group activities are out of phase with other groups connected through WANs or global communications networks. Up to a point this is also a scheduling issue, but it is limited by the fact that there are only a few hours at best in a standard workday of any particular group when all interested parties may be at work simultaneously. This can become a hindering problem, particularly serious when collaboration is taking place between participants in Australasia, Europe, and North America, as is often the case.

When real-time participation in a multimedia videoconference or concurrent engineering design project is not practical or possible by some key individuals, the proceedings of such sessions could be saved for future viewing and comment. The main practical problem is adequate storage of massive multimedia data generated during the sessions, management software, and a means of notifying the desired parties that their attention and comments are urgently required. Theoretically, this is not dissimilar in concept to voicemail or E-mail systems but requires much more voluminous storage and processing capabilities. In practice voicemail or E-mail that are accessed frequently would be the logical means to alert the persons whose inputs are being sought that their participation, albeit belated, is nevertheless urgently required.

Any multimedia store-and-forward system automatically provides the capability for reviewing prior conferencing sessions by one or more participants after the sessions are over. It also provides a basis for developing management reports of specific project activities which can be illustrated or substantiated with pertinent clips of critical videos and statements by the participants by third parties. This suggests the need for specific security measures governing access to such storage and limits on how long full sessions should be stored. The time factor will render such data obsolete relatively quickly, and purging most of such information will also reduce storage costs for these systems. There is no question, however, that the existence and use of such systems will have an impact on the politics and culture of an interactive enterprise of this type.

Such store-and-forward systems also form the basis of imaging systems that acquire, store, retrieve, and transmit images of documents, although most often use proprietary protocols and hardware unsuited for use with more conventional platforms and networks. On the other hand, the existence of ODB systems and applications that are designed to manipulate BLOBs consisting of text, data, voice, graphics, images, and videos provide the basic hardware to accommodate such multimedia store-and-forward facilities. More recently database and client-server suppliers have been enhancing their products with multimedia capabilities. It is quite clear that as time progresses these systems will become important justifiable elements of multimedia networking systems providing storage facilities for time-sensitive applications as well as many other multimedia applications.

Videoconferencing Services

At the end of 1993 there were an estimated 20,000 videoconferencing systems installed in the United States mostly as private links within corporations. A certain percentage of those systems are operated by long-distance telephone carriers and special service organizations that offer videoconferencing to business users or the general public.

Videoconferencing services may proliferate in the future and become as common as fax machines because prices of desktop videoconferencing systems are declining very rapidly and competition is increasing. Some industry observers believe that by 1997 such videoconferencing systems will cost below $1000 per unit and will become an option for PCs such as modems or multimedia capabilities today. (See Fig. 4.7.)

During 1993 the Kinko chain of copy centers announced a videoconferencing service targeted at the traveling businessperson which will eventually become available at over 600 Kinko copy centers throughout North America. This service is designed to offer videoconferencing at about $150 per hour, which is considerably cheaper than the average $250 per hour charged by established videoconferencing service companies.

Specialized Videoconferencing Vendors

There are several groups of videoconferencing equipment, systems, and service vendors, and new ventures are being formed every day. For many years the industry was dominated by specialized videoconferencing suppliers Compression Labs, PictureTel, and Video Telecom (now VTEL) who provided proprietary systems and codecs incompatible with each other. More recently with the introduction of standardization in data transmission, breakthroughs in DSP pricing, and more powerful PCs and workstations, videoconferencing applications have been implemented on desktops, obviating the need for spe-

The two main videoconferencing services available are public-room offerings (A), which are fully equipped facilities rented by the hour, and multipoint conferencing (B), in which providers coordinate transmissions from user-owned codecs located at geographically dispersed sites.

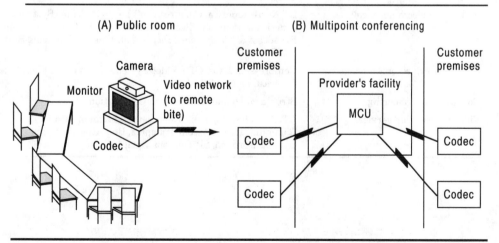

Figure 4.7 Videoconferencing services. (*Source: Reprinted from* Data Communications, *May 21, 1993, p. 69, copyright by McGraw-Hill, Inc., all rights reserved.*)

cial equipment. Rapid growth in networking and the promise of the information superhighway are very powerful incentives for new ventures to develop very competitive low-cost videoconferencing systems. In fact the established videoconferencing suppliers are taking advantage of these market developments by coming out with lower-cost systems and desktop alternatives as well, and most now offer MCUs that allow multiparty videoconferencing between systems with different protocols.

There are basically several product groups that are discernible in the videoconferencing market with the established suppliers active in several market segments. These include the consumer-level videophones, desktop and LAN-based videoconferencing systems, more traditional rollabout products, and complete custom-built boardroom videoconferencing centers. Concurrent engineering and associated groupware systems are a special market segment. In addition, there are videoconferencing services supplied by major long-distance carriers and specialized service organizations and brokers. Table 4.2 identifies these market segments and major vendors active in each market. Names, addresses, and telephone and fax numbers of these firms can be found in the Appendix of this book.

TABLE 4.2   Specialized Videoconferencing Vendors

Product category	Major vendors in this market
Videophones	AT&T, BT North America, Hitachi America
Desktop videoconferencing	AT&T, Compression Labs, Extron Electronics, Fujitsu Industry Networks, Future Labs, IBM, InSoft, Intel, Novell Multimedia, Sharevision Technology, VideoLabs, VTEL
LAN-based videoconferencing	Datapoint, DEC, Hewlett-Packard, Intel, Intervision, Microsoft, Silicon Graphics, Sun Solutions
Rollabout videoconferencing	BT North America, Compression Labs, GPT Video Systems, Hitachi America, Mitsubishi Electronics America, NEC America, Panasonic Broadcast & TV Systems, PictureTel, Videoconferencing Systems, VTEL
Boardroom videoconferencing	Compression Labs, GPT Video Systems, Mitsubishi Electronics America, PictureTel, VTEL
Concurrent engineering	Structural Dynamics Research Corporation
Videoconferencing services	AT&T Global Business Video Services Group, Affinity Communications, Ascom Timeplex, Bell Atlantic, Connexus, MCI Communications, SP Telecom, Sprint Video

5

Sales and Marketing Applications

A number of political, economic, and technological factors are converging during the 1990s leading to continuous intensification of worldwide competition for the conceivable future barring reemergence of a global threat comparable to the Soviet Union of previous years.

Politically, the new world order is seen as three extremely powerful blocks competing for markets within and without their territorial borders. The European Economic Community (EEC) has been in the making for several decades and is finally entering the phase of full economic integration. The United States, Canada, and Mexico are moving toward a more integrated North American Free Trade Association (NAFTA) which may in time encompass most of Latin America. Japan and the manufacturing power bases of South Korea, Taiwan, Hong Kong, and Singapore form the Far Eastern Bloc already flooding the whole world with a continuous choice of new products that have often displaced or even put out of business inefficient manufacturers and suppliers in Europe and North America.

The most significant aspect of the current global geopolitical equation lies in the fact that technology is easily transferred from location to location. It can be adapted rapidly to produce more competitive products anywhere in the world combining financing, specific tax, import and export duty advantages, and just-in-time training of temporary workforces.

In order to survive in such environments businesses must develop superior market intelligence by continuous monitoring, gathering, and evaluating huge quantities of information about shifting customer preferences, changing prices, competitive products, available financing, and distribution channels. In fact the latest theories on global competitiveness focus on timely information as the new and unlimited form of capital. This is true if pertinent information can be captured, and manipulated rapidly into specific knowledge that can be communicated, received, understood, and acted on by those who have the resources and authority to exploit it.

More often than not, marketing executives are already laboring within an "information overload" environment because existing data processing systems are generating more data and information than can be absorbed and understood by average human beings. What's more, this information overload is growing at an exponential rate, particularly with the advent of international data networks, corporate databases, and a proliferation of personal computers expected to number 400 million by the year 2000.

Multimedia, and particularly intelligent multimedia networks and knowledge bases, promise an enhancement of the end-user interface that can go a long way in eliminating some of the information overload by substituting images, animation, voice, and motion video for large volumes of text and numbers. Multimedia solutions appear particularly attractive in developing more effective customer services, including promotions, customization, design, testing, and manufacture of highly customized products to meet specific objectives, in the shortest possible time, and at a profit even in quantities of one if need be. Portable multimedia solutions with instant connectivity to all corporate entities can move interactive enterprises closer to the virtual corporation model as long as companies develop and maintain efficient interactive multimedia networks and communications.

Sales-Force Automation

Increasing numbers of sales representatives are being equipped with microcomputers both within their offices and in the field, and almost from the start such systems have been designed for connectivity with central corporate databases. Currently mobile sales forces have a choice of extremely powerful notebooks, pocket computers, portable CD-ROM players, personal digital assistants (PDAs), and other more specialized electronic devices. Most are equipped with modems and other means of communication.

Originally such field devices were relatively limited in processing and storage capabilities and were used primarily for territory management and reporting. More recently these portable devices have become sufficiently powerful to offer platforms for multimedia applications.

The multimedia approach is considered ideal for promotion and sales of products and services with certain common characteristics. These include product complexity, relatively high unit price, a high degree of customization, extensive information content, or existence as part of a large product line. Numerous financial products and services fall into this category as well as travel tickets and itineraries, pharmaceutical drugs and application therapies, real estate promotions, computer software systems, industrial equipment and systems, and many one-of-a-kind architectural and engineering projects.

Many attempts to automate such applications have been made primarily through the use of expert systems. These resulted in highly interactive systems that attempt to match the needs and expectations of a customer with a description of a specific product or service as well as advice and explanations of why such a choice is justified. While such systems often provide good intelligence and consistency in matching customers with products, they are often considered too boring and technical to be of value for supporting direct sales in the field or the front-office environments.

Multimedia elements in such interactive sales and marketing advisory systems provide a much more engaging and interesting interface and make it possible to deploy such systems for direct use by prospects and clients. Such interactive multimedia applications are expected to become quite popular as sales and marketing organizations and advertising agencies introduce more automation into their functions.

Portable Multimedia Promotions

Portable multimedia promotions and applications are becoming practical and cost-effective solutions as a result of an increasing number of multimedia-capable portable devices reaching the market. These include several categories of portable computers, portable CD-ROM players, portable multimedia peripherals, and PDAs. The most attractive applications are developing in field sales and marketing, customer servicing, diagnostics, just-in-time training, and mobile videoconferencing.

The bulk of such portable multimedia promotions are basically standalone applications because public networks capable of handling interactive multimedia transactions are not widely available. Nevertheless, rapid progress in development of new portable devices and wireless communications networks suggests that portable multimedia promotions are potentially valuable and may soon become cost-effective means of selling many products and services.

Wireless Multimedia Communications

Portable devices such as notebooks and PDAs are often seen as the ideal in sales-force automation, but they also have significant operational restrictions. Even devices with built-in fax and/or modems require phone jacks at fixed locations to provide connectivity. The ideal solution is, of course, wireless communications, but early attempts often required additional specialized equipment and offer very limited bandwidth capabilities. Nevertheless, many industry analysts see wireless communications as the ultimate in sales-force automation once appropriate service networks are in operation.

The market for wireless mobile computing is still emerging, but new technologies and products appear on the scene at a rapid rate. The market still lacks sufficient definition, low cost of entry, and established practicality to attract massive usage, but by the year 2000 about 10 million wireless mobile data network users are expected to exist, according to The Yankee Group, a market research organization specializing in communications. Motorola, which is involved in manufacture of mobile communications equipment, believe that there will be 20 to 26 million two-way wireless data users by the end of the decade, and some analysts go so far as to predict that a wireless market will skyrocket into a $600 billion-per-year industry by the year 2010.

Research studies indicate that about 38 million people are engaged in occupations that require mobile communications of which 4 million are salespeople spending at least 200 days per year on the road. Corporations are increasingly interested in employing roaming workers of that type who can operate from their cars or homes. They not only are more competitive by having access to corporate decision-making databases and expert systems wherever they may be but also reduce the overhead costs of more extensive office environments and working facilities.

Wireless data services solve the problems of portability where plug-in facilities or easy connectivity do not exist as is the case at many airports and some hotels. Strictly speaking, wireless communications provide only local untethered transmission like cellular phones and otherwise rely on proprietary WAN systems to transmit and distribute the data on a long-distance basis. Currently existing wireless services handle voice and data such as E-mail, but it is clear that video transmission will be accommodated in the near future. This is a necessary requirement for wider use of such systems in sales automation since many portable devices are already equipped with multimedia processing capabilities.

Personal Digital Assistants

These are small, lightweight, portable devices that provide an array of communications functions including voice, fax, E-mail, paging, handwriting recognition, and touchscreen controls. Although not strictly multimedia applications since they do not provide interactive audio and video handling capabilities, PDAs compete for a number of mobile markets that are being served by other portable multimedia devices. Developers and vendors of PDAs also expect to add full multimedia capabilities eventually to these products, most of which are much lighter and cheaper than other portable communications solutions. Table 5.1 provides an overview of the initial PDA products and their suppliers.

PDAs are expected to become more valuable portable multimedia devices when broadband networks provide a better communications infrastructure, although some PDAs are being equipped with wireless communications capabilities.

Dataquest projects that revenues from sales of PDAs will soar from $39 million in 1993 to almost $2 billion in 1997. BIS forecasts that by 1998 2.3 million PDAs will be sold, while some highly optimistic estimates by the vendors see more than 100 million PDAs of all kinds in operation by the year 2000.

TABLE 5.1   Personal Digital Assistant (PDA) Products

PDA supplier	Comments
Amstrad (UK)	PDA 600; stylus input organizer type; unit cost $500
Apple Computer	Newton message pad; 20-MHz 32-bit RISC processor; 0.9 lb weight; 21 h of battery power; unit cost $699
AT&T	Personal Communicator (discontinued); EO 440 and 880 models; 20-MHz 32-bit RISC processor; 2.3–4.0 lb weight; 4 h of battery power; unit cost $1999–$3299
Compaq	Mobile Companion; DOS-compatible; Microsoft and Intel collaboration
IBM	Simon; wireless connectivity and graphics; 1 lb weight; unit cost $1000
Sharp	Expert Pad; based on Newton design; 0.9 lb weight; unit cost under $1000
Tandy	Zoomer; developed jointly with Casio; Geoworks GEOS operating system; 1 lb weight; unit cost $699–$899

Interactive Multimedia Advertising

The interactive multimedia advertising came into existence some years ago with the introduction of the so-called electronic brochure. These are diskettes targeted at PC users that promote specific products such as automobiles or services of a company using graphics, animation, and images as well as sound. Now interactive messages can be made available on CD-ROM disks for those with multimedia PCs and CD-ROM players.

A more sophisticated concept involves networked multimedia advertising. In some cases it takes the form of public information kiosks or displays that are networked to present information about an area or event intermixed with advertising messages for associated products and local services. Others are simply multimedia displays without interactivity features, but in all cases these are centrally controlled systems for updating information and advertising messages as rapidly as necessary with the least amount of effort.

The most sophisticated systems consist of a WAN that provides multimedia advertising messages to a professional target audience such as physicians. Physician Computer Network of Laurence Harbor, New Jersey, is such an organization which provides free networked PCs to physicians in return for a commitment to review up to 32 product advertising messages every month and answering a clinically oriented question with each message on an interactive basis. These inputs are collected by the network as statistics on patient demographics and sold to pharmaceutical manufacturers and other market research organizations in the health care industry.

The advertising industry is also positioning itself to take advantage of the interactive TV systems that are being developed by cable TV and telecommunications companies. The concept of a digital superhighway promises to open up a huge home market to interactive multimedia advertising and associated market research on demographics. It is not yet clear what forms these programs will take, but it seems that these will be variants of interactive games, contests, educational programs, and information databanks.

Merchandising Kiosks—Unattended

This is probably the most familiar form of multimedia marketing consisting of a touch screen, a computer, CD-ROM or videodisk, printers, and credit card readers all packaged in an attractive kiosk. These are normally located in high-traffic areas including stores, shopping malls, supermarkets, trade shows, and transportation centers. Kiosks are seen as the fastest-growing multimedia market segment increasing at an average annual growth rate of 120 percent, expected to reach 1.5 million installations by the year 1995.

Kiosks often offer more consistent product information than low-level sales personnel and operate on a 24-h basis without additional cost. They can be easily modified and can be used to demonstrate products, play product-related games, dispense discount coupons, capture prospect names, accept orders, and process credit purchases.

Successful kiosk examples

Florsheim Shoe Stores operate a network of 500 kiosks deployed within the stores. Customers can select shoes by style, size, and color using a touchscreen and following voice instructions. A keyboard is used to capture names and addresses, and a reader accepts major credit cards as payment for shoes that are shipped overnight by UPS (United Parcel Service). Florsheim reports a 20 percent increase in sales through kiosks and by freeing store salespeople to handle customers more expeditiously.

R. Stevens Express is an automated photo machine that is designed to accept film for processing on a 24-h basis at prices 25 percent lower than photo stores. It provides a video photo teller to advise customers on 200 variations of films and processing choices and accepts cash, credit cards, and personal checks as payment.

Similar kiosks are being developed to sell flowers and floral arrangements by displaying specific videos and placing orders automatically with nationwide floral services accepting payment in several forms. Hermann Sporting Goods uses an interactive kiosk to provide information about sporting vacation destinations and offers advertising space to vendors of products and services associated with those sports and locations. The concept of *Multimedia Yellow Pages* is also being developed for deployment at airports, hotels, restaurants, malls, and tourist areas. These offer advertising space with full-motion video, audio, graphics, and animation to merchants in the area who could not afford a kiosk individually.

Desktop Marketing—Attended

This type of interactive multimedia system is usually installed at a specialty store where it assists the sales personnel as well as customers in selecting the best possible product or service. Conceptually these systems are similar to unattended kiosks but are more sophisticated, and the sale is not made automatically through the system. Their content is designed in such a way as to involve the customer and salesperson simultaneously. Furniture stores, hardware chains, automobile showrooms, beauty salons, and home improvement centers are typical sites for this type of application. These systems are also used automatically to gather market research and demographic data whenever they are being accessed by a customer making specific choices among products.

Mannington Resilient Floors, for example, developed a touchscreen system for its dealers that asks the customer what rooms are being redecorated and immediately displays a selection of floor coverings in a typical room. The system automatically keeps track of how many people looked at which patterns and colors and also contains a retail sales training program to assist sales clerks in sharpening their selling skills by focusing on quality and service rather than price.

DesignCenter is a similar merchandising system installed by a Wayerhauser subsidiary in over 100 home centers and hardware stores. It assists sales personnel and customers in designing and displaying models of certain home improvements. More importantly, the system provides a precise bill of materials and pricing required to do the job. It has been credited with over $250 million worth of designs during the first year of operation, and stores in which it has been installed claim that their closing rates for such products have almost tripled.

Similar merchandising systems have been developed by Servistar, a major hardware chain, and some paint manufacturers for making optimal choice of paint products for specific rooms in specific locations. Other systems in this category focus on matching of paint colors, interior decorating, gardening and fertilizer selection, and various projects connected with farming.

A sales improvement system that assists Steelcase personnel in selling office furniture is also a pioneering example of this type of application. The system employs visual showroom analogy and given customer image, financial standing, and performance expectations it displays video alternatives and computes benefits and competitive assessments.

Buyer Workstation Networks

A special networked multimedia application is the buyer workstation installed within purchasing organizations of major department stores and fashion boutiques. The network provides a constantly updated review of the latest fashions, colors, and prices of clothes, shoes, and other items sold in those stores.

These workstations are installed for the convenience of buyers in department stores and provide interactive multimedia presentations on all products offered. They perform the function of a personalized fashion show but in addition are also equipped with interactive features that make it easy for the buyer to order immediately specific items presented in quantities, sizes, and colors desired. Because the workstations are networked, these orders are received by the manufacturer directly and automatically create all the necessary documentation for invoicing and billing the customer. This type of networked multimedia application is a special case of electronic data interchange (EDI) systems that are increasingly used to perform transactions between corporate entities. In businesses such as clothing the need for visual presentation and inspection of products is obvious, but multimedia use with other EDI systems is also a possibility depending on the complexity of the product and transactions involved.

Public Access Information Networks

Public information systems are basically networked multimedia applications sometimes intermixed with advertising messages. Two categories of such systems exist. One includes a network of monitors that displays predetermined sequences of multimedia images and sound, videos, and information but cannot be controlled by the end users.

These information systems are not interactive and time-sensitive in only a very limited way when they provide information about schedules of transportation systems or events. Such systems are controlled from a central location for updating their content, changing messages, and displaying special information. This type of information network can be found at airports, railway and bus stations, arenas, shopping malls, and other public places.

A second networked multimedia information system application is the interactive display booth, or kiosk, to which the public has direct access. In this type of kiosk the end user must activate the process by making a selection on a touchscreen. Such multimedia systems exist in many public locations and also in large office buildings, department stores, and museums and at sporting events and trade shows. In recent years the Olympic Games relied heavily on such multimedia systems to keep audiences informed about rapidly changing events, scores, and developments.

This type of an application may also be used to generate end-user demographics by collecting and accumulating specific data such as the type, interests in a specific location, preferences, and food tastes of a user. How effective such systems are depends on the content, the design of interactive multimedia sequences, and the location. The fact remains that such innocent-looking information systems have the potential to capture and analyze marketing data of considerable value to the promoters and sellers of products and services involved.

Customer Service Networks

The most successful business enterprises are those that are driven by customer needs and desires. According to various studies, customers rate customer service and on-time delivery of products and services as more important in making a sale than price itself. Service reputation also plays a part in the buying process, ranking second only to product performance.

Customer service may take many forms depending on the type of product or service being provided, and it may come into play before and after a sale is made. Before a sale, customer service is basically an information service designed to identify a customer need and match it with the most appropriate product. The more complex the product, the more extensive presales customer service is required to assure the prospects that a particular product will meet their needs.

The problem with this type of service is twofold. The company must maintain a permanent staff of specialists to answer questions and educate customers or relegate the function to the sales force. In either case the personnel is limited in number, seldom available on a 24-h basis, and often customers must line up and wait for service. Research suggests that in competitive environments customers will not wait for service and will switch to another supplier who can provide immediate and satisfactory response and price quotations.

When sales people are used to provide this type of service, there is also the potential of lost sales if the answers do not meet customer expectations. Apart from that, live human responses are not consistent and will vary depending on the person involved and the way that individual feels during a particular day or time of day. Misinformation in order to make a sale can also damage the long-term sales of a vendor more than is generally realized.

Postsales customer service is extremely critical, particularly when complex products are involved that require diagnostics and just-in-time training to provide the proper level of service. This activity has been addressed in previous years with various expert systems and help-desk applications, but it became soon apparent that multimedia knowledge bases provide superior solutions to those that can only manipulate text and numbers. Graphics, animation, and voice commentary are some of the initial media elements used in such systems, but motion video with associated audio is clearly superior and conveys more convincing knowledge more efficiently.

Interactive multimedia knowledge systems focusing on a particular product or service based on local networks provide a solution in various public service areas. Such systems can handle customers in a consistent manner prequalifying prospects for various services that require extensive customization such as insurance policies, financial and investment services, mortgages, and various government services. The most important advantage of these applications is the freeing of sales personnel to close business rather than spend their time evaluating prospects, and dispensing advice.

Multimedia customer service systems can interface directly with customers or provide rapid access to specific knowledge and solutions to corporate service personnel who interpret the knowledge for customers. This approach is necessary in cases of complex technical applications such as servicing of generator turbines in power stations or providing maintenance services for aircraft and helicopters.

Real Estate Presentation Systems

Visual presentations play a major role in real estate promotion and sales which also involve costly visits to inspect properties. The objectives of all such systems are to reduce the frequency and costs of property inspection visits and increase salesperson productivity.

Those systems accept specific input from a prospective buyer who indicates price range, location, size, architectural style, required amenities, distances from schools, places of worship, tennis courts, beaches, and up to several hundred different characteristics. Once these parameters are in the system, the system displays a regional map pinpointing suitable properties. A touchscreen allows selection of images of these properties, and another layer of controls displays images of individual rooms within the house.

Home View Realty Search Service of Needham, Massachusetts, claims to have developed the first such service. Another firm, Electronic Realty Associates of Kansas City, Missouri, also developed a similar system that allows their brokers to show houses anywhere in the world from any of their offices. In Denmark, HomeVision has been developed by the largest real estate chain in the country which gives buyers the chance to browse visually through houses for sale simultaneously collecting vital inputs about family size, location preferences, and financial status. All these are networked multimedia applications with significant potential because such systems can be upgraded to offer virtual-reality presentations that will provide the illusion of walking inside a house and interacting with its structure and objects in it on a real-time basis.

Travel Agency Multimedia Systems

A good example of multimedia potential in the travel industry is SABREvision, a service of American Airlines deployed at over 1000 travel agencies. The system provides maps of destinations, images of hotel lobbies and rooms, and many visual details of local attractions. Coupled with travel reservation systems, this is a prototype of future multimedia systems in this industry.

National Car Rentals Systems and Budget-Rent-A-Car developed interactive multimedia booths that handle car rentals without intervention of counter personnel in about 5 minutes.

In the moving industry, Ryder Touch-TV is an interactive merchandising system providing information about new locations, comparing living costs, developing packing lists, and a telephone hot line directly to Ryder rental offices.

The future of the travel agency business is in question when the potential of interactive TV and similar services targeted at the home are considered. This is an area of considerable interest to various service organizations such as banks and financial services because these companies are among the first to deploy interactive multimedia service networks for their products.

Financial Services Applications

Some of the most promising opportunities for interactive multimedia networks exist within the financial services environments. Most financial products or services meet the characteristics of ideal multimedia applications because they are complex and require extensive customization.

Insurance companies are endlessly seeking methods to reduce the costs of claims processing to a minimum, but at the same time they must make sure that inexperienced agents do not make incorrect conclusions, causing unnecessary payments and losses. The focus is on interactive multimedia systems that facilitate the investigation and determination of liability in accidents dealing primarily with automobiles but clearly suitable for other claim areas. These systems depend on simulations in which agents observe accidents and are shown correct liability assessments. Aetna and Allstate are leading in development of such systems, but many insurance companies are exploring similar and expert-systems-based solutions.

Banks, investment, mortgage, and insurance firms are all developing or exploring interactive multimedia solutions to reduce waiting lines, provide more competitive services, and train their employees. Some banks are considering an expansion of the automatic teller machine (ATM) concept to include other financial services in a "branch of the future" in which insurance, investments, mutual funds, and travel services can be dispensed to the customers without personal assistance of bank employees.

The investment industry is already using video broadcasts from Financial News Network (FNN), Cable News Network (CNN), Reuter, Knight-Ridder, or Dow Jones News within a broker's workstation. There is also development of a voice-controlled trading workstation jointly sponsored by Citibank, First Boston, Goldman Sachs, Morgan Stanley, Salomon Brothers, and Shearson.

Virtual-Reality Simulations Potential

Virtual reality (VR) is a simulation technology that includes computer-generated images to create an illusion—sometimes known as *cyberspace*—that the user is interfacing with a real environment. VR offers the capability to "walk around" simulated models of various structures such as office buildings, automobiles, engineering prototypes, or complex molecules.

In marketing one potential application is simulation of product displays in a shop or department store allowing customers to visualize new furniture in their homes or design a whole kitchen. In the business world VR may come into use as a simulation of complex multimedia LANs and WANs to determine their performance before actually undertaking their development.

VR technology is claimed to have a broad range of practical applications ranging from teleconferencing and architectural design, to collaborative computing and entertainment. Find/SVP predicted that the market for VR applications will explode from $5 million in 1991 to $575 million in 1995. Frost & Sullivan estimates put it at $84 million in 1992, reaching $1 billion by 1997.

VR has considerable potential as a multimedia simulation tool in conjunction with collaborative computing and concurrent engineering, but the demands for massive high-speed data processing required to support VR simulations limit it to well-financed operations with access to extensive computing power. Companies like Boeing, Chrysler, Fujitsu, Glaxo, IBM, Matsushita, NEC, Sony, and Toshiba are the early adopters already exploring commercial VR systems and applications.

Major Products and Service Suppliers

Most sales and marketing applications are based on four types of computer products. These include desktop PCs and workstations, portable computers, CD-ROM players, and PDAs of various types. Many PC manufacturers supply multimedia-capable platforms based on 486 processors with 66-MHz speeds, built-in speakers, and CD-ROM drives that are suitable for interactive multimedia networking.

Among portable computer vendors, Dolch and Toshiba were the pioneers of multimedia platforms, although these units are relatively heavy and expensive. More attractive are the latest notebook models such as IBM's ThinkPad, Apple's PowerBook, or NEC Ultralite, which have the power to handle multimedia but require external speakers and CD-ROM drives to operate. Two examples of very compact notebooks specifically designed to handle multimedia with built-in CD-ROMs are produced by Panasonic and Scenario Systems.

Specific portable CD-ROM and CD-I players are manufactured by Sony and Philips, but these devices are low-priced platforms with limited processing and networking capabilities. A multiplayer device based on the 3DO design is also manufactured by Matsushita and competes with basic CD-ROM players at the low end.

As for PDAs, there are several vendors with relatively limited devices on the market. These include Amstrad, Apple, Compaq, IBM, Sharp, and Tandy. These companies have either developed or are in various stages of development despite relative lack of enthusiasm for PDAs on the market at present.

Industry analysts believe that by the mid-1990s there will be at least 18 vendors offering various PDA products at very competitive prices.

In the area of wireless communications the most interesting are providers of wireless WAN services through mobile radio, cellular phone, or satellite networks. While these are not capable of providing broadband transmission facilities at present, these services are the prototypes of future networks that may offer faster transmission speeds and make interactive wireless multimedia a reality.

6

Multimedia Training and Information Networks

One of the oldest and most popular uses of interactive multimedia technology is in corporate training. Originally known as *computer-based training* (CBT), it has been in use for over 20 years in various forms. As hardware costs came down, desktop computers proliferated, and integrated CBT authoring tools became widely available, CBT applications evolved from mainframe-based textual systems through interactive videodisk into full networked multimedia applications.

The use of interactive multimedia in corporate training applications is particularly valuable when processes, products, and services involved are complex, frequently upgraded and modified, and require continuous customer support. The trend toward the virtual corporation operating model which may require specific skills at short notice creates an additional demand for networked multimedia training systems.

Major training areas where multimedia technology has been applied include management skills, industrial factory floor training, information technology products, medical and health care, government services, and military weapons systems operation and maintenance.

Multimedia training applications are usually developed by a team effort with conceptual and design inputs from subject experts, educational psychologists, and creative designers. Advanced multimedia authoring systems make it relatively easy to design and deploy interactive multimedia training without extensive computer or programming skills.

Many such programs are delivered in form of CD-ROM disks, and there are numerous training "titles" available developed by specialized organizations in many industries. There are also CD-ROM LANs in which case a variety of training programs stored on special CD-ROM players are available to users who can download a training course of choice for use at their own desktop. A survey in *CBT Directions,* a multimedia training magazine, revealed that already at the beginning of 1992, 21 percent of all CBT applications were networked on the basis of OS/2 or UNIX platforms.

Training on Demand

With the advent of interactive enterprises wishing to engage in virtual corporate marketing opportunities, training on demand will become a necessity. Training on demand is not unlike just-in-time training already practiced in manufacturing environments. The difference lies in courseware which is known ahead of time in the case of manufacturing processes but is unpredictable within a temporary virtual corporation.

Training on demand can be easily accommodated within the real-time interactive multimedia communications environments that will prevail in the virtual corporate mode. By definition, a *virtual corporation* is a combination of several interactive enterprises united temporarily by a common marketing objective. There is absolutely no reason why one or more of such interactive enterprises could not be a specialized multimedia training service providing precisely the type of training required to achieve the objective of the virtual corporation. The capability to operate within a virtual corporation environment will automatically imply sufficient networking bandwidth capabilities to handle multiuser conferencing, and multimedia training on demand will be no problem to such organizations.

There are also opportunities for specialized organizations to provide such services, and in all probability firms that already produce specialized training courseware will supply that market by providing their services on a networked basis.

Distributed Corporate Training

Multimedia technology has been used in corporate training for some years, particularly in form of CBT systems with interactive videodisk platforms. Traditionally multimedia CBT applications have been designed for standalone platforms and distributed to end users by means of laser disks, CD-ROMs, or compressed software on magnetic media. This is an adequate and cost-effective method of distribution for many training applications that do not require frequent updating.

There are many industries, however, where continuous training is a fact of life and where product or technology changes are very frequent. In such cases networking of workstations and PCs for delivery of multimedia training programs becomes very important. It is most often justified by the fact that if it can be delivered through a network, it saves considerable time and expense of assembling employees from distant locations at specific training sites. This is particularly true in the case of financial service and consulting firms with numerous locations where employees must be kept up to date about the latest complex and varied competitive products offered by their organizations. Other organizations with high personnel turnover such as in the hotel industry or retailing business may also benefit from networked multimedia training facilities available whenever required at all locations.

Because multimedia training is not time-sensitive, distribution to remote locations by CD-ROM or other media is often more cost-effective. On the other hand, certain just-in-time training programs may have to be delivered rapidly at an industrial location that experiences an emergency. This can be accomplished through workstations on a LAN with a specific server dedicated to the storage of multimedia training programs. On the other hand, corporations are expanding the bandwidth of their LANs and WANs in order to handle increasing traffic and to introduce multimedia conferencing. In such cases corporate training programs can be distributed using those infrastructures at a marginal additional cost and certainly more efficiently than using other methods of distribution. (See Fig. 6.1.)

Distance learning

Distance learning makes it possible to mix text, video, audio, and other multimedia objects while freeing an instructor from having to be in the same location as students.

Training on demand

Hypermedia documents running on videoservers will enable users to learn at their own pace. Interactive learning system and training on demand may revolutionize education.

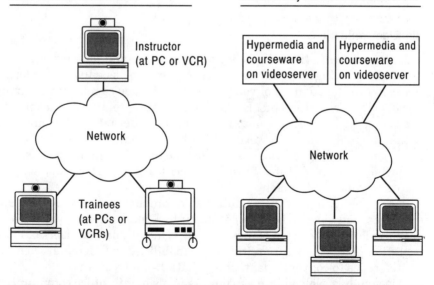

Figure 6.1 Networked multimedia training applications. (*Source: Reprinted from* Data Communications, *September 1992, p. 88, copyright by McGraw-Hill, Inc., all rights reserved.*)

Just-in-Time Training

Just-in-time (JIT) training is primarily designed to assist factory workers who experience difficulties with machine tools or a particular component on the assembly line. In many cases JIT training applications developed from initial expert systems designed to handle such situations. However, introduction of multimedia elements such as graphics, animation, and video into JIT training systems enhance their effectiveness by presenting real demonstrations.

A JIT training system logically questions workers, collects their responses, and automatically determines the solution. It then provides specific knowledge explaining the situation and demonstrating a procedure to fix the problem or get the system going again. It is a variant of the diagnostic system, but its focus is more on the training function to minimize the occurrence of similar stoppage in the future.

In the virtual corporation operating environment there are also numerous situations that may arise requiring JIT training for workers engaged in other activities. Theoretically a virtual corporation brings together interactive enterprises each of which is specialized in the tasks it needs to accomplish. However, virtual corporations are also operating under tremendous time constraints and within very competitive environments. As a result, there may be many instances when an expert interactive enterprise is not available. In such cases JIT training of available pools of workers may be the best solution. Whether the JIT programs are generated in house or by outside organizations is not really important. What is critical is the ability to transmit such JIT training to the locations where it is needed in the most efficient way.

Multimedia Electronic Mail

Electronic mail (E-mail) is basically a store-and-forward system which is rapidly growing in popularity as a corporate messaging and information service. With the addition of voice, image, and video capabilities, E-mail becomes videomail or multimedia E-mail and offers a relatively inexpensive way of exchanging business communications of considerably better quality and value than textual E-mail. However, E-mail or videomail do not offer time-sensitive transmission capabilities nor secure means of communications. As a result, videomail is not suitable for mission-critical communications nor as a substitute for videoconferencing.

A whole range of new multimedia product enhancements allows voice, images, and video to be incorporated into E-mail systems which use LANs and WANs for transmission of messages. It is inevitable that users of PC and workstation platforms with multimedia capabilities will demand such products in preference to conventional text and data E-mail services.

E-mail has evolved into a primary means of interoffice communications, and many companies are now integrating different E-mail systems that have been installed in various organizations over the years. Specialized gateways and backbones are also available to perform such tasks. Originally, E-mail messages were almost exclusively short text items, but now E-mail is increasingly used to send all types of files including documents, spreadsheets, graphics, voice annotations, and even video clips. As a result, the cost of integrating E-mail systems with multimedia capabilities may be very high relative to multimedia videoconferencing alternatives that can always be used in videomail mode once suitable and adequate storage is available.

On the other hand, existing E-mail systems present a huge market potential for multimedia capabilities. E-mail systems are very popular, and their use is growing rapidly. At the end of 1993 over 26 million E-mail users were estimated in the United States alone. Forecasters predict that by 1996, 77 million E-mail users will exist worldwide.

The interoperable documents

E-mail is only one communications mode that is used in corporations alongside voicemail and fax services, all of which use the telephone networks for connectivity. The trend now is to justify the cost of videomail systems by replacing all such services by a single server system that can handle E-mail, voicemail, and fax services. With such a videomail or multimedia mail system the user can receive messages in whatever form they arrive or choose a special format of received messages regardless of what format was used for the original input. Such interoperable documents could be sent out in one form and received in another. Some products of this type are already on the market, and more will undoubtedly appear as information superhighway segments come into operation. Figure 6.2 illustrates the basic concept of an interoperable videomail system.

E-mail is an established networked application, and multimedia is primarily a transformative technology, which makes it very well suited to enhance existing E-mail systems. E-mail is not a time-sensitive application; therefore, it is relatively easy to add voice and video to E-mail messages as long as it is realized that very significant additional storage will be required to handle multimedia elements. The addition of these multimedia elements to E-mail results in a much more powerful communications tool and leads the way into compound document mail systems that eventually can manipulate such documents in transit. Software products for manipulating E-mail transactions are already on the market, but success of these applications depends on sufficient bandwidth between end users and storage areas or servers on their LANs.

When E-mail is viewed as a subset of a broader messaging function in a corporation, it is clear that it will be enhanced with multimedia capabilities. E-mail traffic often includes directory information, EDI electronic forms, network management and monitoring information, account reporting, reliability reports, E-mail filters, schedules, calendars, interactive group workflow, remote user meetings, and other applications. Most of these can be enhanced with images, sound, or video to improve collaborative applications. As a result, several E-mail products already offer multimedia dataset handling capabilities

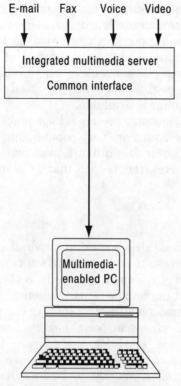

E-mail   Fax   Voice   Video

Integrated multimedia server

Common interface

Multimedia-
enabled PC

User chooses format of received message

Figure 6.2   Interoperable videomail system.

including audio, images, and video as well as JPEG and MPEG compression algorithms.

Corporate Information Services

This type of multimedia application is basically a network of strategically located kiosks within the corporate environment. These kiosks are similar to public access information services in concept, but the content of information is company-specific and designed to answer many questions that employees normally pose to their supervisors or human resources departments.

Corporate information services provide current information about the status of the enterprise, its top management team, location of company facilities, corporate product lines, R&D programs, representative clients and new contracts, and company performance.

A separate information stream may handle more employee-specific information such as safety procedures, available benefits, company training programs, new job openings, organized social events, housing assistance, health care support, and other areas pertaining to the welfare and advancement of the individual worker.

These information systems save the time of middle management in answering and explaining various issues within a corporation. Many restructuring programs significantly reduced the number of middle managers and put greater demands on the time of those remaining. Under such circumstances corporate information services fill in an important but not revenue-producing function. They also can be updated from a single location, which assures accuracy and timely dissemination of corporate information. The same information can be broadcast through video distribution systems to employees equipped with PCs that can handle multimedia applications. The advantage of kiosk-based information services lies in the fact that these can service other employees who do not have PCs or access to such equipment.

Multimedia File Sharing

Multimedia file sharing is not so much an application as a necessary store-and-forward capability of networking systems for handling such applications as multimedia E-mail, video distribution systems, imaging systems, concurrent engineering, and multimedia conferencing.

The need to share multimedia content in networked environments poses unique challenges and requires new multimedia server solutions to optimize the support for sharing multimedia data. Basically multimedia file sharing calls for multimedia servers that must control very large volumes of continuous datastreams, multimedia databases that have the capacity to store such BLOBs of multimedia data, and networking and communications facilities that can distribute such datastreams which may be accessed by end users in real time.

Multimedia data may exist in digital or analog form; therefore, multimedia servers must be able to control both data types. Multimedia file sharing may involve analog servers that control storage devices such as laser disks or TV tuners and digital servers which control the multimedia data itself, retrieving it from files of databases.

Some of the object databases (ODBs) that can store and manipulate text, data, voice, graphics, and video provide suitable storage facilities for development of multimedia file-sharing infrastructures. Numerous multimedia applications that are not time-sensitive will reside on special servers whose files will be shared by many users.

Video Distribution Systems

Video distribution systems are passive multimedia applications designed to allow the end user to receive video transmissions from a variety of analog sources such as cable TV, broadcast TV, closed-circuit corporate TV, VCRs, and videodisk players. Video distribution systems are in effect mini cable TV systems using PCs or windows on PC screens as displays. Because these are analog videos they can only be displayed on PC screens alongside digital content

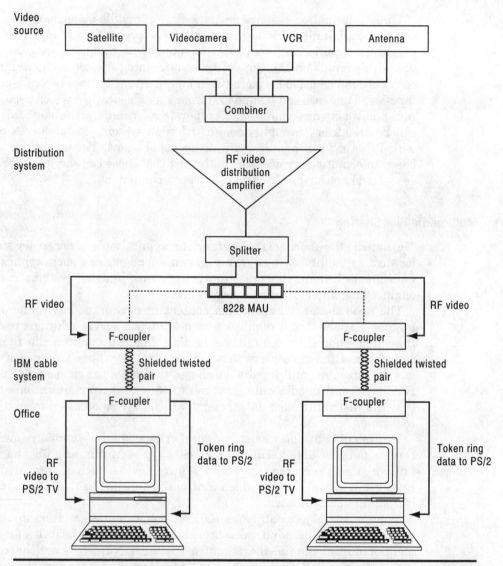

Figure 6.3 Anatomy of a video distribution system (MAU—media access unit). (*Source: IBM sales brochures.*)

but cannot be manipulated or combined with it. An example of a video distribution system is outlined in Fig. 6.3.

Video distribution systems may exploit existing LANs for transmission of signals but require specific hardware boards or adapters in PCs to provide TV windows on the screens. If desktops are not multimedia-ready, speakers or earphones are also required as an additional device. Several PC hardware and peripherals suppliers offer specific products in this category.

TV broadcasts

This application basically consists of distribution of imported TV signals directly to desktop PCs. The analog video signals are displayed on the whole screen or in a window, and control software simulates TV controls which are comparable with those found on conventional TV sets.

In the corporate environments TV broadcasting on a PC is used in a number of ways. The previously mentioned corporate information service is one obvious application, but it requires in-house or contract video production capabilities to produce, distribute, and update specialized corporate videos.

Another common use of PCTV (personal computer TV) is the display of news channels such as CNN, C-Span, and FNN and other specialized broadcasts in medical or computer fields to corporate managers and workers who must stay in constant touch with world or industry developments. Financial services executives, such as stockbrokers, are a good example of businesspeople who use complex networked workstations for trading and benefit from real-time windows with a choice of TV news channels to support their advisory and decision-making activities.

Training programs

A very large proportion of corporate training programs as well as information services are not interactive but consist of conventional videotapes. These are used individually or in conjunction with classroom lectures or presentations in conference rooms. The existence of a video distribution system provides an alternative method for delivering this type of training program directly from a corporate headend broadcasting center to suitably equipped PCs or videoconference facilities.

Video library access

Corporations maintain archives and libraries of videotapes and videodisks that are made by corporate staffs, contract video production firms, and specific video publishers. These archives include historical information, research data, previous training programs, product descriptions and promotions, and records of corporate events and activities of its executives. Once a video distribution system exists, such archival videos can be cataloged on-line and made accessible to corporate and outside users on demand.

Surveillance, Detection, and Intelligence

When the sources of video input are strategically located, videocameras become video monitoring systems for surveillance of buildings, manufacturing process monitoring, hospital nursing stations, and store monitoring. Such systems are not interactive but rather collect and record video information which is normally monitored by human security guards. If video distribution systems exist in a particular corporate environment, such surveillance systems could provide input if necessary for distribution to other areas of the company. It is not inconceivable that such systems could be developed to the point where detection of certain activity patterns even unattended will trigger preventive action like sounding alarms, locking strategic passages, and simultaneously broadcasting such information with pertinent video segments to appropriate executives for further action.

Imaging Systems

Until recently imaging systems were the exclusive domain of specialized system vendors who supplied expensive proprietary products. These can scan, store, and process images of documents but operate with specialized protocols and cannot be easily integrated with predominant networking environments.

These traditional imaging practices are changing with a trend toward integrated approaches based on client-server architectures. This means a departure from the complete proprietary solutions toward image management software products that can manipulate images stored in specialized databases and catalog programs.

The imaging standalone market is declining rapidly, and imaging is now seen as another data type that should be handled by networked PCs and client-server systems. Major PC application vendors are developing products which will incorporate imaging as a component of compound document processing software. The objective of PC-based imaging is the ability to scan and digitize pertinent items and images from any document or publication and attach them to E-mail, spreadsheets, or other productivity software products. In order to assist the imaging user on the PC, a whole new range of products is now becoming available that provide image catalogs and databases. Many provide image managers and thumbnail images for editing and selection purposes and are also useful to multimedia application developers.

Traditional imaging has certain drawbacks because it uses proprietary interfaces and protocols which may prevent digitized documents from being processed by any other workstations except those designed to operate with the imaging system itself.

Imaging functions include scanning, digitizing, compression, storage, manipulation, display, and printing of documents, and these are similar in development and operation of multimedia applications. Since networking protocols and compression schemes are now largely standardized and delivered on PCs and workstations using LANs and WANs, document imaging can be regarded as a function of the multimedia communications environment within a corporation. Companies may be streamlining their imaging systems to perform seamlessly with collaborative multimedia projects being put in place.

Major Vendors of Products and Services

Multimedia training and information networks are usually integration projects which involve corporate employees from different departments and vendors of various hardware and software products and tools. These include authoring systems, E-mail software, and imaging systems, as well as specialized multimedia databases, filing, cataloging, and server systems. In many instances major vendors of computer hardware and software such as IBM, DEC, Sun Microsystems, Hewlett-Packard, Microsoft, Oracle, or Sybase undertake turnkey projects of this type. Some major system integration consulting organizations are becoming increasingly involved in these markets, with Andersen Consulting being a clear leader in the field.

In the specific area of multimedia CBT tools and systems development there are a large number of relatively small vendors who offer authoring systems or corporate training design and implementation services. Others provide ready-made training courseware on practically all business and industrial subjects which must be integrated into a corporate networking infrastructure. These include companies like AimTech, Allen Communication, CBT Systems USA, Computer Teaching Corporation, Claris, Comsell, Concurrent Technology Developers, Effective Communications Arts, Gain Technology, Global Information Systems Technology, Macromedia, Multimedia Learning, Ntergaid, Performax, Spectrum, Synesis, Technology Applications Group, and Wicat. A good source with up-to-date information about all CBT related suppliers is the annual *CBT Buyers Guide* published by Weingarten Publications of Boston (MA).

In the E-mail segment Lotus Notes, Microsoft Windows for Workgroups, and WordPerfect Office are the well-known suppliers of E-mail products. However, there are many more vendors with network-based E-mail products, a number of which can handle multimedia data types such as audio, images, and video as well as JPEG and MPEG compression algorithms. They include Applix, Beyond, Digital Equipment, Enterprise Solutions, Frontier Technologies, Innsoft International, Lotus Development, Microsoft, Verimation, and WordPerfect.

In the imaging arena major traditional vendors include companies such as Excalibur Technologies, FileNet, IBM ImagePlus Systems, Optika Imaging Systems, Wang Laboratories, and Westbrook Technologies. There are also many more PC-based image cataloging and database vendors. They include companies like Electronic Imagery, GTE, IEV International, Lenel Systems, Nikon Electronic Imaging, PowerSoft, Synoptics, and Videotex Systems.

7

Health Care Industry Networks

Multimedia networking has a major role to play in the restructuring of the health care industry, which in this instance includes the physicians, hospitals and clinics, testing laboratories, pharmaceutical companies, medical equipment vendors, a vast R&D establishment, and medical insurance providers. Once a patient is under treatment or hospitalized, most of those entities come into play and there is often intense interaction, consultation, and conferencing between all parties involved with a constant need to demonstrate, authorize, and document symptoms, procedures, and therapies.

A recent study by Arthur D. Little, a leading consulting organization, concluded that the health care industry spends unnecessarily about $36 billion annually because it lacks certain information technologies. Specifically these include bedside computer terminals, videoconferencing between teams of physicians, home health terminals, and diagnostic systems. In fact it is believed that health care is 5 to 10 years behind most of U.S. industry in use of information technology and still relies mostly on paper records and manual filing systems.

Multimedia networks are ideal solutions in such cases, providing pertinent images to physicians, hospitals, and medical records almost simultaneously. This reduces significantly the time required for delivery of such documents and images and enhances the quality but reduces the cost of health care delivered to patients. Medical imaging networks that can handle complete patient records within a region are seen as one of the most productive concepts for networked multimedia applications.

The Clinton Administration health reform proposals suggest that restructuring in this industry should rely heavily on the private-sector deployment of information technology to create more effective health care delivery. On the other hand, hospitals, many of which lack the cash resources, are not seen as the most likely initiators of change, which implies expenditures on networking hardware and software systems. As a result, the field appears to be wide open for third parties to provide interactive multimedia networking services including capturing, storage, transmission, interpretation, and updating of all the information in such systems.

Networks and specialized databases are seen as the most important information technologies which are required to implement the proposed restructuring of the health care industry. One projection suggests that this will increase hospital expenditures for information systems from about $4.5 billion in 1992 to $6.7 billion by 1996.

Restructuring of the Health Care Industry

The Clinton Administration health care proposals are a major factor in the heightened interest in specialized medical multimedia networks because these are seen as a key component in cutting costs and streamlining delivery of medical services. A major stumbling block if the information exchange between health care providers and insurers is to be enhanced is the lack of standards in patient records, which must include current medical, diagnostic, and insurance information.

The structure of the Clinton health care reform plan is widely debated and will undoubtedly continue to change as time progresses. However, what is not questioned is the fact that multimedia networking technologies provide the appropriate infrastructure to put it into effect. The proposed National Health Information Infrastructure is to combine EDI, with image, audio, and video into regional networks and WANs providing nationwide multimedia communications between all parties concerned.

Another aspect which is not clear is how different health care units and individuals can connect with such a network to engage in effective interaction with each other. There is a very significant obstacle in the existing traditional ways of administering health care which will require massive changes in attitudes and procedures. Currently health care providers such as hospitals have already automated some procedures including diagnostic X rays which are taken at different locations and sent to a central radiology depository. However, there are no incentives in developing automated links between such isolated electronic depositories and other interested parties or insurance companies.

The administrative processes between health care providers and insurers are seen as the major initial area where restructuring of the industry must take place to lower costs and improve delivery. Next in line is the exchange of patient medical records between physicians and specialists and ultimately the development of diagnostic assistance networks.

The restructuring process involves numerous applications which must be standardized and matched with the most effective technologies to result in cost reductions and improvement of services. These applications include patient enrollments in health care programs, eligibility issues, authorizations, claim

TABLE 7.1   Elements of a Health Care Network

Health care function	Interactive communications requirements
Patient enrollment	On-line data communications and storage; past history may involve image processing
Eligibility determination	Expert system evaluation, E-mail, voicemail, data retrieval from external sources, and data communications
Authorization	Rule-driven advisory systems, extensive messaging in all forms, data retrieval and communications with many different internal and external sources
Claim submissions	Extensive image processing potential for still and video data transmissions with expert diagnostic and multiparty videoconferencing systems with internal and numerous external sources
Coordination of benefits	On-line data communications, voicemail, E-mail, and data storage
Specialist opinions	Appointment scheduling, image and data processing, voicemail, E-mail, and data communications and storage; multiparty videoconferencing potential
Hospital admissions	On-line data communications, image processing potential, bedside terminals on LANs, alarm systems, extensive data processing and storage, real-time monitoring
Laboratory testing	On-line data communications retrieval and storage with numerous internal and external sources
Test results interpretation	Image and video transmission processing; multiparty videoconferencing potential; extensive voicemail, E-mail, forms processing, diagnostic systems, and data retrieval
Patient records	Image and video processing potential; voicemail, E-mail, expert diagnostics, and data retrieval systems
Medical supplies	Data retrieval and communications, just-in-time training, E-mail
Prescriptions	E-mail, voicemail, pharmacological drug interaction systems, FDA records, storage and retrieval

submissions, benefit coordination, appointments with specialists, hospital admissions, laboratory tests and result interpretations, drug prescriptions, medical supplies, and patient records (Table 7.1). All these must be introduced in electronic form into the networks with real-time interactivity and image processing to be effective.

Everpresent Need for Visual Demonstrations

In the operating room there is always a need to consult physicians who may not be able to attend the procedures for financial or geographic reasons. Associated with such consultations is the need to visually demonstrate the state of the patient and provide real-time images of radiological X rays, CAT (computer-assisted tomographic) scans, electrocardiograms, and surgical procedures being undertaken. This suggests that a real-time multipoint videoconferencing or visualization system is the best solution in such instances.

In brain surgery, for example, the traditional approach is to examine X rays and CAT scans of the patient's head, which during the operation is immobilized in a metal frame. The surgeon must work around this awkward obstacle, imagining the exact location of a tumor before drilling and removing part of the skull. Use of an interactive 3-D visualization system removes that spatial guesswork out of neurosurgery because the system uses digitized CAT scan data to create an image of the brain. The surgeon can call up interior 3-D images of the patient's brain by manipulating a probe on the patient's head and locate the tumor and best drilling position precisely.

To be effective in a given community, a health care multimedia network must include all the operating units in the area and provide real-time connectivity between all such units on a random basis. These normally include hospitals, clinics, HMOs (health maintenance organizations), referring physicians, specialists, medical records, and insurance providers, as well as imaging sources such as X rays, CAT scans, MRIs (magnetic resonance images), and past image depositories. This presents a significant challenge because such units are usually scattered throughout the community and must be connected with MAN or WAN networks with sufficient transmission capacity to provide multiple transmissions simultaneously between numerous parties. Figure 7.1 illustrates such a network based on Media Broadband Services (MBS) offered by NYNEX.

Special Imaging Requirements

The necessity to demonstrate and consult in real time is probably critical in cases of surgery, but there is also a constant need to exchange relatively complex images of X rays, CAT scans, and MRI documents as well as visualizations of patients and their conditions. This means very high resolutions for workstation displays to make them effective for on-screen diagnostic functions. It also means transmission of very massive multimedia data between various parties often on a real-time basis.

Three-dimensional medical visualizations may require rendering up to 30 megabytes (Mbytes) of data instantly to make them realistic and useful. As a result, the links between the different medical units in a community should provide bandwidth capacity up to 100 Mbps to provide acceptable performance and even more capacity if several transactions of that type must be accommodated at the same time.

Patient Monitoring Networks

A lot of technology is being used in hospitals to monitor patients, but more often than not these are local systems whose output is relayed to other parties through manual means. There are many opportunities to enhance health care if such systems could be integrated through networks with the various services that are involved. One aspect that would be valuable to the physicians and specialists is the ability to provide them with the image of the patient wherever

MBS applies both to the Metropolitan Area Network (MAN) environment where remote physical locations need to be connected as well as to the Local Area Network (LAN) environment where image creation devices are integrated with display and storage capabilities.

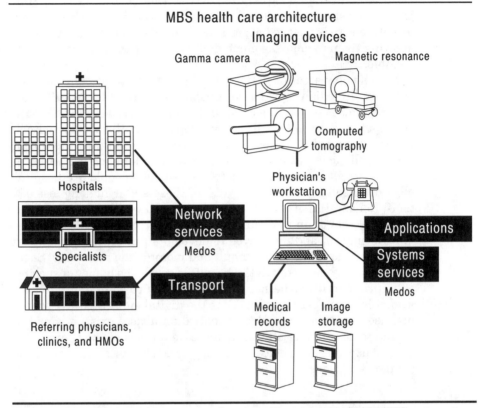

Figure 7.1   Example of a health care network system. (*Copyright 1994 by NYNEX.*)

they may be and automatically raise the alarms in emergencies. If such systems can be connected to specific expert systems, automatic patient monitoring would extend to special actions when incorrect medicines that could trigger unwanted adverse reactions are prescribed or administered.

As part of a larger interactive multimedia network, such patient monitoring systems would be accessed from a workstation within each patient's room by physicians, nurses, and even patients with special training, instruction, and demonstration sessions available at the push of a button when required. The stations would also serve as inputs to all information collected about the patient on a real-time basis, thereby obviating the need to keep numerous paper records and costly duplication of such efforts when transmitting information from one party to another.

A further benefit of such networks would be its ability to monitor patients in the home, thereby reducing the need to stay in the hospital and cutting down on the cost of health care. A multimedia network of that type is already in use at the New England Medical Center for monitoring children with cancer. The interactive system instructs parents how to change dressings and administer blood tests to their sick children. When the blood counts reach a dangerous level, the system automatically alerts the physicians in charge of the case. The incentive here is to save hospital costs but also ease the traumas of patients by allowing them to enjoy as much quality time with their families at home as is possible.

A whole new area for patient monitoring and treatment appears to exist in using multimedia programs to rehabilitate patients who suffered brain damage. The images, videos, sound, and graphics associated with multimedia can act as stimuli to help a brain develop new pathways to route information around damaged neurons. These multimedia computer therapies are highly individualistic in nature and must be tailored for each particular patient.

Government estimates indicate that 500,000 people are hospitalized annually with head injuries and 90,000 of those suffer lifelong loss of brain functions. Rehabilitation is a time-consuming and very costly procedure sometimes reaching costs of up to $4 million per patient. As a result, there is a very significant market and incentives to develop an interactive multimedia therapy designed to generate such stimuli and monitor the responses. Such a program could be easily delivered in the hospitals as well as homes of such patients if an appropriate multimedia network infrastructure were in existence with access available from every household as the digital superhighway concept seems to promise. Once the network and central databases exist, it would be relatively simple to accumulate statistics and case studies of the therapy that could be used to expedite the choice of therapy and rehabilitation by new patients with similar conditions.

Multimedia Diagnostic Networks

The Clinton health care reform proposals among others advance the concept of community-based information networks that would allow physicians to exchange clinical data about their patients on a random basis wherever they may be. This type of network requires image processing as its basis as well as videomail or videoconferencing and direct access to diagnostics and patient records at the same time. The theory is that any physician visiting any particular patient can use any workstation to access any patient records and can be called to respond to any situation with any patient that may have triggered alarms within the system.

An extension of this concept is to install workstations within homes of other patients that do not need to be hospitalized but must be monitored constantly by physicians. There is no clear vision as to who would administer network-oriented home care, but one proposal suggests that physicians provide that service under contract to insurers at specific fees while third parties design and maintain the interactive multimedia networking infrastructures that would be required. The availability of access from each household to the digital super-highway would greatly facilitate the implementation of such schemes.

Medical Training Networks

The health care industry experiences a very rapid rate of change in procedures and instrumentation as a result of technological developments, discovery of new therapeutic mechanisms, and the introduction of new drugs into the marketplace. As a result, there is a constant need to disseminate new information and methodologies to all medical practitioners in a timely fashion. In addition, there is a constant demand for demonstrating real surgical procedures to medical students.

Because surgery is an extremely rigorous occupation, surgeons in training are required to accumulate thousands of hours more of training than other medical practitioners. This is in addition to keeping up with the new techniques and procedures that all physicians must absorb at all times. One opportunity in this area is the development of interactive multimedia systems that create virtual environments simulating real surgical operations. Such systems are already in development, and it is argued that they will speed up the training process of surgical residents in particular who would be able to practice and try to duplicate operations performed by the most accomplished surgeons of the world.

Currently there are many standalone interactive multimedia training programs available on CD-ROMs and other media for use by medical students, nurses, and other medical workers. These courses vary in length from 1 or 2 hours to hundreds of hours and include numerous tests designed to discover how well the student is absorbing the material. Although such courseware is helpful and useful in demonstrating therapies and various medical procedures, it does not provide the real-time effect or demonstrate proper behavior in unexpected emergency situations as well as actual videoconferencing systems can. What is important, however, is the fact that large numbers of medical workers are already familiar with interactive multimedia training programs, and making their availability via networks at any time and any place would help significantly in keeping those workers up to date with the least amount of time.

Pharmaceutical Drug Promotions

Presentation of new drugs and associated administration procedures to sophisticated medical audiences requires use of interactive multimedia systems. These are developed for promotion of specific drugs and are designed to capture the attention of physicians at specialized conferences and symposia where competing products are often also presented. Some of the most innovative multimedia applications have been developed for these competitive environments, including specialized challenges and games related to medical topics and drug performance.

At least one interactive multimedia network is in operation which promotes specific drugs directly to physicians. Physician Computer Network (PCN) is supported by pharmaceutical companies which provide information about their drugs and specific therapies for their effective use. The network also provides physicians with valuable productivity software designed specifically for use in automation of their mandatory reporting and documentation procedures.

The introduction of such a network into the offices of individual physicians is not an easy task and has to overcome a number of natural obstacles. One is the general dislike of computers and fear of losing control and also a chronic lack of time to get involved with such matters. As an incentive, PCN provides the computers to physicians free of charge in return for a commitment that they will spend a certain amount of time going through the drug advertising sections and will answer a few questions pertaining to the material they have seen. The objective of such a network is to collect demographic data about reactions of physicians in the field to specific new drugs and their preferences to use them in specific therapies. It is not inconceivable that in the future physicians will receive free multimedia access equipment from various sources and will be able to obtain drug information and training programs from numerous on-line services specializing in medical promotions of this type. The challenge remains in capturing their attention and desire to use such services, and developers are looking to interactive multimedia to make these systems attractive and interesting enough for this purpose.

Physician Symposia and Conferences

Interactive multimedia presentations in form of kiosks or special booths and large screens have been in use during medical symposia and conferences for some time. This activity is an extension of the use of interactive multimedia in promotion and advertising of drugs and new medical equipment and procedures. The lessons learned from these presentations suggest that it is necessary to pay special attention to the content and format of these presentations in order to capture the attention of attending physicians. As a result, many of these interactive presentations have been developed in form of challenging game-like programs that pitch users against real or imaginary experts in solving specific medical situations and require them to draw on all their existing knowledge and skills to participate. Some of these programs are connected with various prizes and special deals offered to winners of specific challenges.

Future Trends

Industry observers look into the future of health care automation by focusing on the capabilities of a physician's workstation that they believe will eventually exist in all doctors' offices. These workstations will have very high-resolution displays and graphical performance capabilities because they will allow on-screen diagnostics of pathology, and will have the capabilities to perform real-time exchanges with other physicians across the country or around the world if necessary.

Such networked multimedia workstations will be able to display very clear X-ray images with facilities to point, scale up, and focus on specific areas of interest and create prints for documentation if necessary. Availability to process EKG (electrocardiographic) signals and retrieve immediately patient records will be an automatic function in all such workstations connected to specific databases. There will also be high-resolution cameras and scanners attached to these workstations that will allow a physician to excise a tissue from a patient, digitize it instantly, and send the image for evaluation to one or more specialists for immediate discussion and diagnosis.

Suppliers of Medical Multimedia Networks and Products

A videoconferencing network that allows physicians to consult with each other during surgical procedures and operations, known as VideoCare System 1000, is already being offered by the United Medical Network Corporation of Minneapolis (MN). The elements of this videoconferencing system include standard equipment from PictureTel and special arrangements with MCI Network Services. The key to this network is relatively low cost of communications which is independent of mileage and time of day and is negotiated with the carrier by the network provider. The Health Care Information Network is another organization that provides an exchange of patient medical data and administrative information.

A voice-driven reporting system integrating patient case information and data has been developed by Kurzweil AI, a leading voice recognition organization. It is known as VoiceMED and integrates emergency medicine, with radiology, surgical pathology, and other specialties into a hospital information system.

Interactive Multimedia Consumer Markets

The attention being given to interactivity and multimedia in the media results from a perceived potential for massive networked multimedia consumer markets associated with the proposals and the development of the digital superhighway concepts. This publicity is responsible for feverish investment as well as merger and acquisition activity among potential participants within the telecommunications, cable TV, and entertainment industries.

This frenzy of activity and interest within the consumer markets is also responsible for development of new multimedia-capable hardware, software, and service products by traditional hardware and software developers who perceive numerous business opportunities as purveyors of enabling technologies and systems to the interactive multimedia consumer markets.

Nevertheless, these consumer markets are very risky at present because there is little knowledge or agreement about the type of networked multimedia services and products that might be of interest to consumers in the long run. Despite many tests of various interactive multimedia applications, the profitability of these services is still in question, but several major joint ventures designed to test and justify various forms of interactive multimedia consumer entertainment and information services are already under way.

Numerous applications are being developed and tested with different end-user audiences in various localities. The results of these tests, particularly with regard to delivery methods and end-user preferences, are of great value to all multimedia applications developers.

These networked multimedia developments in the consumer market are of considerable significance because the massive size of this market provides an incentive to many investors, large and small. As a result, hardware and software products facilitating implementation of interactive networked multimedia applications in consumer as well as business areas may come to market sooner and at much more competitive prices than would otherwise be the case.

Perceptions of Massive New Markets

Most of the enthusiasm about interactive multimedia in general and in the consumer markets in particular stems from a perceived potential for automating and replacing massive existing markets that are estimated at about $120 billion annually in the United States alone. Currently, these markets are fragmented into 10 major areas including catalog shopping; broadcast and cable TV advertising; consumer video; information services; music recordings in form of records, tapes, and CDs; movie theaters; videogames; electronic messaging; and videotelephone and videoconferencing (Fig. 8.1).

These markets are dominated by the numerous catalog shopping organizations that alone account for over $50 billion yearly. Broadcast advertising is the second largest segment, estimated at about $27 billion annually followed by home video sales which at $7 billion per year already surpassed previously long-established markets for music recordings, movies, and videogame industries.

Most of these market segments are already involved in some form of electronic service and product delivery, but automation levels vary greatly and

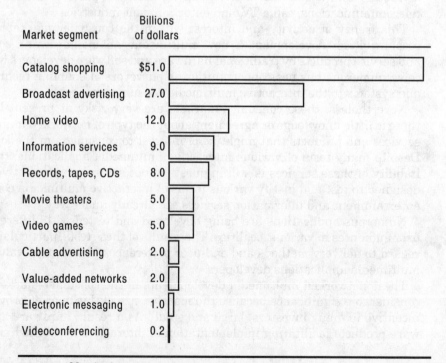

Market segment	Billions of dollars
Catalog shopping	$51.0
Broadcast advertising	27.0
Home video	12.0
Information services	9.0
Records, tapes, CDs	8.0
Movie theaters	5.0
Video games	5.0
Cable advertising	2.0
Value-added networks	2.0
Electronic messaging	1.0
Videoconferencing	0.2

Figure 8.1 Massive interactive consumer markets.

interactivity is very rudimentary and limited when it exists at all. Now the concept of the digital superhighway and the availability of interactive multimedia technologies promise to create profound changes in all these market segments. These changes will favor large organizations that control the digital networking infrastructures. They include telephone and cable TV companies as well as major information and entertainment content developers and archivists. Those are the film studios, and a whole gamut of publishers of books, tapes, CDs, CD-ROMs, videogames, and other products of this category.

Those companies believe that they are in a unique position to control a whole new industry combining entertainment, consumer electronics, and computing to develop new services that will be delivered directly through existing TV sets and rapidly increasing populations of PCs in the homes. Many of these companies understand that going interactive is the only way their businesses will survive in the future. This situation also explains the merger and acquisition frenzy that has overtaken this environment during the last few years.

The major issues now confronting the promoters of interactive multimedia services on the digital superhighway are to discover what consumers really want on their TV sets, which services will be profitable, and what is the most effective method of delivery into the home. In order to answer some of these questions, a number of telephone, cable TV, software, and hardware companies have already formed various partnerships designed to test consumer acceptance of interactive TV concepts and services. Table 8.1 outlines the major efforts already in different stages of realization.

Initial market research based on interactive multimedia network trials already under way suggests that the most attractive services will be those that interactively offer entertainment, related transactional shopping, ability to communicate with others electronically while engaged in the process, and instant availability of additional information on the subject through hypermedia functions.

This type of testing will continue for some time in various parts of the country, with different population segments, presenting a wide range of information, entertainment, and educational services at different price levels. This is necessary because the digital superhighway will eventually have the capacity to provide up to 500 TV channels into the home, and it is necessary to discover the most appealing offerings to compete effectively in such an environment.

The promoters are betting on the insatiable thirst of consumers for all forms of entertainment. They see interactive multimedia as a new vehicle to provide more of it, to more people, more often, and highly customized for individual tastes, times of engagement, and ability to pay. Most importantly such capabilities will yield much more precise demographics, which will provide a basis for targeting specific audiences for additional products and services.

Many of these market segments are also part of an even larger rapidly growing indoor and outdoor recreation and entertainment industry which also includes toys, sporting goods, gambling, amusement parks, pleasure craft and vehicles, live entertainment, spectator sports, and other forms of entertain-

TABLE 8.1   Major Interactive TV Pilot Projects

Service or Pilot	Details of service, sponsors, and suppliers
Bell Atlantic	Pilot project involves 2000 homes in Alexandria, Arlington, and Fairfax counties in VA; uses Oracle Media Server
EON, Inc.	Began life as TV Answer based on Hewlett-Packard computers and radio transmission; provides games, home shopping, and bill-paying services; pilot operates in Fairfax county, VA
Full Service Network (FSN)	Most ambitious interactive TV piot to date by Time Warner Cable using Silicon Graphics videoservers, AT&T switching, and Scientific-Atlanta control units; service introduced to 4000 households in Orlando, FL, area; offers movies-on-demand, electronic mall shopping, and video-phone and videoconferencing in the future is being planned
Interactive Network	Backed by NBC, A.C. Nielsen, United Artists, and Cablevision, this pilot uses FM radio transmission; system allows viewers to play TV game shows such as "Jeopardy" and "Wheel of Fortune" and select sporting events
Main Street	GTE is testing a cable-based interactive TV system designed to acquire data about shopping, travel, games, local information, and referencing services usage patterns
One Touch	Viacom and AT&T joint venture testing video-on-demand services in Littleton, CO, and Castro Valley, CA; initial tests are in 1000 homes, to be expanded to 4000
Prodigy	The IBM/Sears joint venture is enhancing its service by adding interactive TV features using Jerrold/General Instrument control units for TV sets
US West	Pilot project in 2500 Omaha, NB, homes planned to expand to 60,000 households; US West has access to FSN technology at Time Warner because it holds 25% of that company's stock
Videotron	A Canadian pilot in Montreal, Quebec, tested interactive advertising in 4 languages during TV broadcasts; allows selection of video angles and information about players in sports events; small audiences tested to date

ment and recreation. These combined represented consumer spending of $341 billion in 1993 and as a whole are seen by some analysts as the new economic engine taking over from the huge defense industry expenditures of the cold war era. Most of those markets also present opportunities for use of interactive multimedia, while some provide the sources of content for many multimedia products. Gambling presents a special case because it is illegal to offer such a service over interactive TV, but there are powerful lobbies that are seeking to ease those restrictions in the future. Interactive multiuser videogames over the digital superhighway are also seen as a very promising new business where prizes may take the form of products, services, and instant celebrity status throughout the network where such an activity took place.

The MCC "First Cities" Initiative

The Microelectronics and Computer Technology Corporation (MCC) research consortium organized the "First Cities" research project in October 1992 with the explicit purpose to explore the feasibility of creating a seamless environment for the spontaneous use of integrated, interactive multimedia services in the home, in the community, or on the move. The objective is to explore the potential for networked multimedia information and entertainment products and services. This research is focused on specific services including multimedia conferencing, interactive games, entertainment on demand, home shopping and transaction services, customized multimedia information, distance learning, and health care.

The group originally focused its research on potential for interactive high-definition TV (HDTV) services. As it became clear that HDTV will not impact the consumer markets until the turn of the century, First Cities shifted its focus and began investigating the market potential of interactive services that would exploit existing TV technologies. It is now exploring the potential of networked multimedia applications delivered through a variety of facilities such as coaxial cable, copper wire, fiberoptic cable, cellular phone, wireless, and satellite systems.

First Cities organized several test sites around the country to measure market response and evaluate an architecture that will accommodate numerous technology platforms and existing telecommunications infrastructures. Corporate members of the group include Apple, Bellcore, Bieber-Taki Associates, COMSAT Video Enterprises, Corning, Eastman Kodak, Kaleida Labs, MCC, North American Philips, Southwestern Bell Technology Resources, Sutter Bay Associates, Tandem Computers, and US West. It is interesting to note that many of those organizations are also involved directly in joint ventures among themselves actually testing these concepts in the real world.

Entertainment on Demand

The concept of entertainment on demand is relatively simple. It involves a network which links a large database of entertainment videos or movies that can be accessed by anyone with a TV cabled into the system and equipped with a control box that allows selection of specific programs at any time and provides the billing information to the service organization.

This form of multimedia video distribution already exists in hotels where the TV sets in rooms often provide interactive services including recent movies on demand. Video-on-demand is the initial growth segment in hotels which can offer guests choices of new movies released well before these reach the video-stores. Spectradyne, which is the largest provider of video-on-demand Guest Choice services, has equipment installed in over 765,000 hotel rooms in 2600 properties worldwide. LodgeNet Entertainment, OnCommand Video, and Guest Serve are major competing suppliers.

Spectradyne recently allied itself with Electronic Data Systems (EDS) for developing the first digital video-on-demand system for hotels. It will also provide additional interactive multimedia applications such as special events, videoconferencing, and education.

Interactive TV

The concept of interactive TV is relatively simple. TV sets that already exist in the homes are equipped with a set-top control box that is actually a powerful computer which decompresses digital video data for display. It also provides a link for the user to interact with the service supplier. Industry observers predict that within 5 years interactive TV could involve about 10 percent of over 93 million TV households in the United States.

The user can request a movie, a TV program, a game, or access to home shopping whenever desired without being concerned about TV schedules or time of day. By the same token, a viewer can respond to TV broadcasts and commercials or use the system for other interactive services that will eventually sprout in the information superhighway.

There are significant benefits to the promoters and advertisers who use interactive TV. News organizations will be able to poll hundreds of thousands of viewers in virtual real time as events unfold without bothering anyone on the phone. Advertisers will be able to obtain immediate orders for their products or services presented in the commercials.

Interactive TV, although technically feasible, is still in experimental stages. Interactive TV delivery centers on the existing TV set in the home and the special set-top control box but the full specifications for effective devices are yet to be understood. The transmission of interactive multimedia signals varies according to the system in use. Interactive TV systems can use cable, telephone, cellular, radio, and satellite facilities for transmitting the programs, and several tests are under way.

The most obvious are interactive TV systems that provide digital interaction via fiberoptic cable lines which can offer up to 500 channels. These systems provide video transmission signal over high-speed optical fiber links to local telephone switching centers where it is switched to a subscriber line compressed under MPEG standard and received through a decompression box connected to the TV set. How much intelligence is put into the converter box determines the degree of interactivity that these systems will have and the range of features and services. What is becoming more important is not how interactivity gets to the home but what is offered that will capture the attention of the end user.

Other interactive TV systems use continuous cellular links and satellite transmissions for two-way interaction or transmit control data via radio and receive viewer response by telephone. Figure 8.2 illustrates the principles of these alternatives.

Although interactive TV service is technically demonstrable, some unanswered operational issues must be resolved before a successful service formula

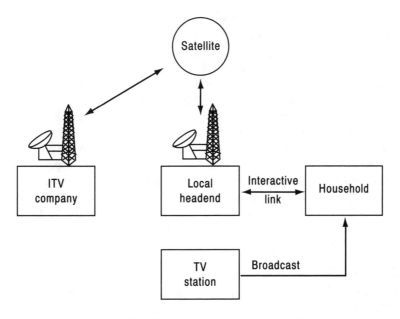

(*a*) Continuous interactive connection

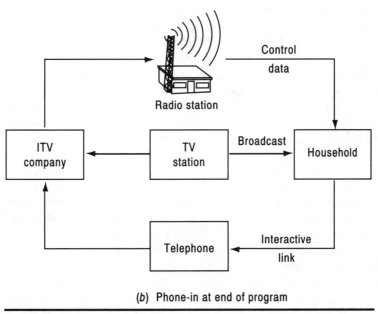

(*b*) Phone-in at end of program

Figure 8.2   Interactive TV alternatives.

is clearly discernible. It is still not certain that existing networks and their associated hardware and software are sufficiently robust to handle thousands of requests for movies, games, broadcasts, or shopping simultaneously. A major issue in such systems is not only delivery of the service to the viewer but also the accounting of who consumed what, for how long, and at what price and the preparation of an accurate record and billing statement.

An associated issue is sorting out the wealth of statistics about viewers, who requested what, when, with what frequency, and for how long. This information is the basis for more precise demographics that interactive TV is promising to its sponsors as well as a target for action by a host of privacy protection agencies whose actions will undoubtedly tend to limit the value of such demographics. The ultimate question, of course, once all technological problems are taken care of, is to determine who is willing to pay how much and for what in order to structure the most profitable operations.

Home Shopping Networks

Home shopping is a consumer application that is seen by many interactive multimedia services promoters as with a huge business potential. The prototypes for this activity are the very successful TV shopping channels where users must interact by telephone to buy products demonstrated on the screen. Interactive home shopping offers an improvement on two counts. It permits immediate response without bothering with the telephone through the TV screen and the control unit and offers viewers control of what products and services they want to see. As a result, viewers do not have to waste time looking at many products of no interest to them but can focus on products which they are interested in buying immediately.

The introduction of interactive TV for entertainment services automatically provides the mechanism for truly interactive home shopping services. In fact, most of the pilot projects now taking place include home shopping as one of the first services.

In effect, home shopping applications turn the TV set into a merchandising kiosk in the home with which complete business transactions can be completed. The competition which will develop among merchandisers will then depend not so much on access to the viewers but rather on the speed with which the vendors are able to deliver the goods to their customers.

The prospect of 500 channels appears to offer an opportunity for many marketing enterprises to develop highly targeted promotions based on viewer responses. While some industry observers are concerned about what content to deliver on so many channels, it must be remembered that the catalog shopping industry includes hundreds of companies large and small, each of which would be delighted to have an interactive channel of its own. This $50 billion annual industry will have to change to sell its products through these interactive channels because otherwise it will not be able to survive a host of new and specialized home shopping firms that will take advantage of this new marketing infrastructure.

Electronic Magazines

Electronic magazines exist on-line or in CD-ROM format and in most cases are basically an electronic form of existing paper (hard-copy) publications. In CD-ROM format these magazines are used with PCs equipped with CD-ROM drives or CD-ROM players on a standalone basis and are usually operated with hypertext or hypermedia features. Although these are standalone products, it is conceivable that a number of such CD-ROMs can be stored in a CD-ROM server or jukebox which could be accessed through a LAN by a number of users. A CD-ROM magazine is comparable to a database, but it may be enhanced with audio, video, and images if necessary. Any interactivity in these is limited to the hypertext mechanism which is basically an information retrieval system.

The on-line electronic magazine are basically original paper magazines whose text has been scanned and stored in digital form on a distribution network such as *America Online* or *Prodigy*. Interactivity is such cases is quite primitive, limited to keyword searching and hypertext in most cases, and multimedia content may not even exist. However, some electronic magazines in this format like *Time* and *Kiplinger's Personal Finance* offer on-line conversations with subscribers through an E-mail capability. Since E-mail is developing as a multimedia facility, these magazines may evolve into truly interactive multimedia magazine services in the future.

The electronic magazine is still in developmental stages, although it is clear that the existence of a digital superhighway could drastically accelerate the process. During the American Magazine Conference in Orlando (FL) in October 1993 multimedia publishing and distribution topics represented a third of all sessions. Magazine publishers are clearly exploring the potential of this concept with considerable enthusiasm.

Real Estate Search Services

These are specialized interactive multimedia networks that are built around the concept of a visual househunting service. These systems require inputs from a prospective buyer who indicates the price range and location of preference. In addition, you can specify up to several hundred different characteristics such as architectural style, size of living area, and nearness to schools, places of worship, tennis courts, public transport, beaches, recreation areas, and similar. These parameters are entered into the system which responds with a regional map identifying the location of suitable properties that meet most of the requirements.

A touchscreen brings up a selection of multiple pictures of these properties, and a deeper layer displays images of individual rooms within the house. An alternative is to couple these systems with an expert system that considers the financial situation of the prospective buyer and suggests a financing program including a choice of mortgage providers.

Other versions of these systems include floor plans and touch screens that permit the simulation of a guided tour of the house with synchronized voice commentary. The obvious next step is to provide virtual-reality experiences of a property by an illusion of walking inside the house and interacting with its walls and furniture. A virtual kitchen along those lines has already been developed in a Tokyo department store, and similar programs are being developed for simulating office space and furniture arrangements.

The objectives of these home search networks are to reduce the number of visits by clients to inspect properties. This procedure is both costly and time-consuming to the purchaser and realtor because on the average it takes about 15 visits before a typical buyer makes a decision to buy. Initial results using interactive multimedia networks of this type suggest that such visits can be reduced by about 50 percent. The access to these systems is still controlled at the viewing centers of real estate agencies, but it is inevitable that once interactive TV becomes a reality, this application is a prime candidate for direct access from the home systems.

Home View Realty Search Service of Needham, Massachusetts, claims to have developed the first such service. Another firm, Electronic Realty Associates of Kansas City, Missouri, also developed a similar system that permits their brokers to show houses anywhere in the world from any of their offices. In Denmark, HomeVision is a similar network implemented by that country's largest real estate firm. These are all networked multimedia applications with significant potential because such systems can be upgraded to offer virtual-reality presentations and can be directed for delivery in the home.

Multimedia Educational Networks

In the consumer markets practically all the cable TV, telephone companies, and other interactive TV players either plan or offer multimedia education of one type or another. It stands to reason that with 500 channels of the digital superhighway there will be many dedicated to interactive training, learning, and education in the home.

A whole new segment of edutainment programs is also aimed at these consumer markets. These are basically educational programs that have an element of games built in in order to keep the attention of the user and develop an interest in the subject matter. The popular "Carmen Sandiego" program is often given as an example of such products.

It is not inconceivable that organizations like vocational and correspondence schools will develop educational courseware that will be networked to subscribers with interactive TV or PCs in their homes. This also suggests the possibility of training in the home administered by employers or contractors.

Existing and Potential Service Suppliers

Major players are telephone and cable companies that are basically locked in a struggle to determine who will control the transmission of video, voice, and data to the home. Much is at stake because such services offer continuous cash streams that will increase significantly as time progresses. They include telephone companies like Bell Atlantic, BellSouth, NYNEX, Southwestern Bell, and US West as major players. Among cable companies the leading actors include Time Warner, Tele-Communications, Viacom, Cox Enterprises, and QVC.

Companies like AT&T, GTE, Hewlett-Packard, IBM, MCI Communications, Microsoft, Oracle, Silicon Graphics, and Sprint also play a role in these new services. However, those companies are regarded more as technology suppliers without experience in consumer entertainment and information markets. This does not preclude the possibility that some of these firms, notably Microsoft, will not attempt to expand into interactive multimedia services in the consumer markets.

What is most likely to happen is that many highly specialized software or publishing firms will develop interactive multimedia products and use the networks operated by the major service companies to try their luck in developing new businesses on a worldwide basis.

Multimedia Networks
Bandwidth Requirements

Interactive multimedia communications basically means two-way processing of multimedia data between the users and a variety of sources and destinations. This presents a number of issues related to the type of multimedia objects being transmitted, bandwidth capacity of the communications facilities, and the nature and number of interacting parties. All this is further complicated when interactive enterprises collaborate temporarily to form virtual corporations whose markets and scope of operations are basically unpredictable. This means that any number of any interactive enterprises should be able to participate at any time in extensive multiparty multimedia conferences and manipulate a variety of multimedia objects in an efficient and timely fashion.

Such multimedia objects may range from a simple text retrieval to intricate voice-annotated changes in a complex 3-D visualization model of particle physics. The range of bandwidth involved is truly staggering. A typical database text retrieval requires about 1 Kbps, whereas a complex visualization needs an 800 Mbps throughput, which is almost a million times greater.

In between those extremes there is a whole gamut of multimedia objects such as text, graphics, audio, and video, each of which requires a different amount of bandwidth for timely transmission within computers and across the networks depending on the type of application of which they are a part. The most demanding of those multimedia objects are high-quality videos with synchronized audio that must be transmitted interactively and in real time. As a result, audio and video transmissions are practical only when compression schemes are incorporated within the transmission process.

This variety of multimedia objects must be transported over private and public networks and data transmission facilities consisting of various types of analog and digital links whose configurations offer different bandwidth capabilities. These may range from 10 Kbps of the traditional telephone lines all the way to the 1.2-Gbps capacity of high-speed ATM cell relay services. Bandwidth capacities also depend on the nature of connecting media, which may range from copper wire, through coaxial cable and optical fiber to photonic networks of the future. Switching and interfacing mechanisms present varying bandwidth capacities that must also be taken into account in multimedia networking design.

The nature and number of interactive multimedia network collaborators significantly influence the bandwidth requirements and performance of such systems. In interactive enterprises it is inevitable that network traffic will consist of random users accessing client-servers, multimedia databases, and engaging in videoconferencing. Many of these applications involve many users simultaneously in real time on a multipoint-to-multipoint basis.

The design of such collaborative multimedia systems must consider bandwidth requirements of the most demanding traffic patterns as the upper limit. Such requirements can then be matched against the available overall bandwidth of all the facilities involved in order to determine whether the resulting transmission latencies and delays are tolerable to the end users of the system. This is not an easy task in cases of unpredictable virtual corporate operations, and for this reason there seems to be considerable opportunity for the emergence of specialized multimedia transmission services leasing bandwidth on demand transmission facilities for as long as these are needed.

The interactive enterprises must be prepared to transmit multimedia objects with different bandwidth requirements across networks and hardware units with varying capabilities. The tempting solution is to make all paths of multimedia traffic capable of handling and storing the most demanding interactive real-time conferencing transmissions for the largest number of users. There are two reasons why this may not be the best approach. In interactive enterprises collaborative efforts are hard to predict and the costs of a network with such maximum throughputs would be prohibitive.

The overall network design must also consider the economics of its interactive multimedia communications requirements. In order to minimize cost and optimize performance, network segments where workers do not require maximum bandwidth capacity must be identified and their connections scaled down accordingly. On the other hand, because of increasingly unstable corporate working environments, all networking facilities should be chosen keeping in mind the need to protect existing investments and ability to expand bandwidth capacities if necessary in the future with the least amount of cost and effort.

In the final analysis enterprisewide multimedia capabilities must provide *real-time conferencing,* which implies high-bandwidth networking facilities all the way to the PC on the desktop. It also means the linking of products including multimedia enhancements of the PC, specialized videoservers for analog and digital data, videoconferencing systems, multiport control units (MCUs), and internetworking products. More importantly, in order to collaborate in virtual corporate efforts despite bandwidth variations, the interactive multimedia user should be able to link seamlessly with workers in other organizations

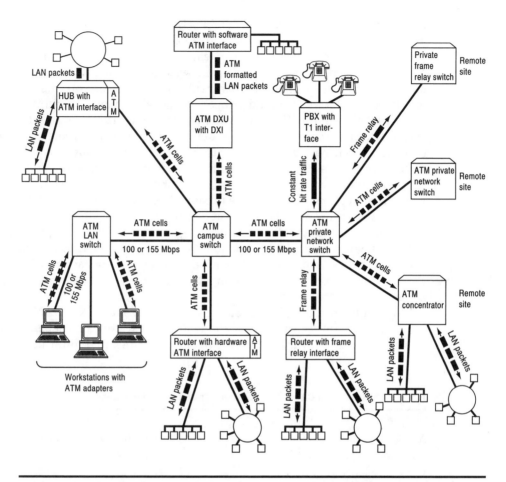

Figure 9.1   Complexities of enterprisewide multimedia network. (*Source: Reprinted from* Data Communications, *October 1993, p. 68, copyright by McGraw-Hill, Inc., all rights reserved.*)

through switched public circuits and cable TV networks and the forthcoming digital superhighway. The complexities of an enterprisewide multimedia network are illustrated schematically in Fig. 9.1.

A Question of Bandwidth and Quality

Any multimedia object can be transmitted through any existing line or network, but unless the bandwidth of the facility is sufficient, the quality of transmission will suffer. This applies particularly to display and transmission of video which is time-sensitive and requires continuous high-bandwidth transmission capacity. There is always a tradeoff between quality of transmission and bandwidth of the medium, and depending on the application, different levels of transmission quality are acceptable by the users. Size of image, frame rate, and compression are the variables that determine the quality of a multimedia video transmission within a given bandwidth facility.

Bandwidth of a transmission link can be defined as the amount of data that a particular medium is able to transmit within a given unit of time. Depending on the capacity of the medium, bandwidth is usually measured in bits per second (bps), kilobits per second (Kbps), or megabits per second (Mbps). In the case of LANs bandwidth is equivalent to throughput of a particular line or a specific path through a network. Many existing transmission facilities are defined by their bandwidth transmission capability as shown in Table 9.1.

Bandwidth of a transmission link depends on the types of cables and connections which form part of the facility. These vary from traditional copper wires used in telephone networks to coaxial cables typical in cable TV and optical fibers that are now used for high-speed transmission facilities and backbone networks. Bandwidth capacity ranges of basic connecting media used in networks will vary depending on quality and purity of materials used and size of conductors, but each has its specific limitations. These factors will also affect the quality of transmissions indirectly. Table 9.2 lists the basic capability ranges of transmission media.

When multimedia traffic is routed across different LANs, WANs, and internetworking facilities, the overall bandwidth of a system also depends on transmission speeds of various switching mechanisms and interfaces that are

TABLE 9.1   Bandwidth Capabilities of Basic Transmission Facilities

Type of transmission facility	Bandwidth range
Analog telephone lines (POTS)	2.4, 9.6, 19.2, 38.4, and 57.6 Kbps
X.25 packet switching	9.6–56 Kbps
VSAT Satellite Communications	15–56 Kbps
Switched 56	Single-channel 56 Kbps
Basic-rate ISDN	128 or 144 Kbps
Frame relay	56 Kbps to 1.54 Mbps
T-1 and fractional T-1	384 Kbps to 1.54 Mbps
Primary rate ISDN, T-1	1.54 Mbps to 24×64 Kbps
T-2 leased lines	6.312 Mbps
Ethernet and token ring LANs	10–16 Mbps
SMDS (scalable)	1.17–34 Mbps
T-3 leased lines	46 Mbps
Fast Ethernet	100 Mbps
FDDI	100 Mbps
T-4 leased lines	273 Mbps
Broadband-ISDN	150–1200 Mbps
SONET standard	51.84–4976 Mbps

TABLE 9.2   Bandwidth Potential of Transmission Media

Media	Distance range, mi	Transmission rate ranges, Mbps
Twisted-wire pair	0.6–6	1–10 (difficult)
Coaxial cable	0.6–6	10–100
Microwave links	0.6–6	10–100
Optical fiber	6–60	100–1000
Photonic networks	n/a	1000–10,000

involved in routing the data. This becomes quite a challenge when multimedia communications require transmission across many facilities with differing equipment under varying load conditions which also change with time. This is particularly true when interactive enterprises combine into a virtual corporation and must communicate through facilities of differing standards of quality.

Levels of Complexity

Existing networking infrastructures require additional hardware and software in order to provide multimedia capabilities on the desktops, and this increases the levels of complexity of such systems. Enterprisewide videoconferencing draws on a wide range of technologies linking conventional videoconferencing systems and MCUs with desktop multimedia and a whole new class of videoservers.

In order to operate as conferencing stations desktop PCs must be upgraded to include video capture and playback adapters, cameras, speakers, headphones, and microphones. CD-ROMs are usually part of any multimedia platform used for playback of off-line multimedia applications although even some on-line projects may require local access to specific CD-ROM databases. In the not-too-distant future writable optical disks will also be in wider use as depositories for massive multimedia transmissions and records of videoconferencing conversations.

Existing client-servers which use popular network operating systems like Netware, Banyan Vines, IBM LAN Manager, or Appleshare can be upgraded and new versions of such operating systems are being released with multimedia processing capabilities built in. At the desktop operating system level, multimedia extensions which were originally available are now part and parcel of the latest versions of DOS, OS/2, Windows, Windows NT, System 7, and UNIX products. As such, these support time-sensitive data objects such as digital videos, audio, and animation and provide hooks into the enabling hardware that handles video capture, playback, and editing. The additional layers of software and their relationships to existing networking structures are shown schematically in Fig. 9.2.

Users are now finding that many productivity software products such as spreadsheets, word processors, E-mail, databases, and similar packages

On the client side of a video application, multimedia extensions tie together the user interface and network operating system. On the server side, the videoserver software is layered on top of the network operating system.

Client

Audiovisual applications
(Lotus Notes, Microsoft Word for Macintosh, Release 5.1a, etc.)
Establish user interface

Multimedia extensions
(Microsoft Video for Windows, Apple QuickTime, IBM Multimedia Extensions to OS/2 Presentation Manager)

Support time-variable data objects such as digital video, animation, sound. Also provide hooks to boards that handle video capture, playback, editing

Network operating system client software (Novell Netware, Banyan Vines, Appleshare)	**Network operating system** (DOS, OS/2, Macintosh System 7)

LAN protocol
(Appletalk, IPX, TCP/IP, Vines, proprietary)

Network adapter
(ATM, Ethernet, FDDI, token ring)

Videoserver

Videoserver software
Furnishes audiovisual synchronization, ensures rapid access to storage and network resources

Network operating system

LAN protocol

Network adapter

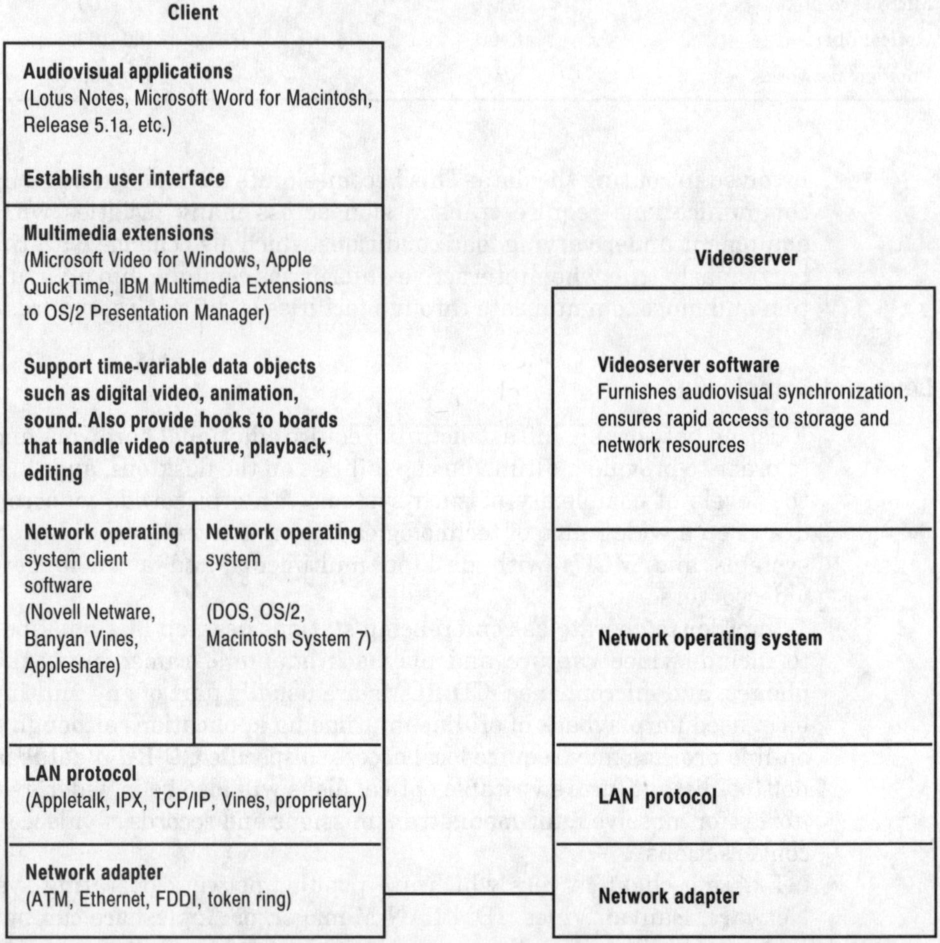

Figure 9.2 Multimedia networking enabling interfaces. (*Source: Reprinted from* Data Communications, *February 1993, p. 63, copyright by McGraw-Hill, Inc., all rights reserved.*)

include links and interfaces to multimedia extensions in operating systems such as Video for Windows, IBM's Multimedia Extensions for OS/2 Presentation Manager, Apple's QuickTime, or the latest UNIX versions with FluentLinks. On the videoserver side, specialized software provides audiovisual synchronization and ensures rapid access to storage and network resources.

Network adapters provide linkage with LANs such as Ethernet, token ring, or FDDI. Logical connectivity between users and videoservers is established by LAN protocols such as IPX, TCP/IP, or Vines, while proprietary protocols support streaming videos.

The videoserver is the key to enterprisewide multimedia capability. It is a specialized client-server dedicated to handle video traffic which must be integrated into existing LAN infrastructures. Depending on the range of activities envisaged for video within an interactive enterprise, the videoservers may become complex clusters of high-performance workstations. Their function is to frame compressed video signals for transmission across a LAN using a mix of video compression devices to deliver the required variable-bandwidth transmissions. Videoservers also perform the functions of linking the interactive enterprise to circuit-switching services and public videoconferencing systems and are capable of converting analog TV and VHS (video home system) signals to compressed digital video.

Matching Multimedia Traffic with Bandwidth Capacity

The key to understanding all multimedia communications issues is the question of matching multimedia traffic with bandwidth capacity of the networks and their individual components. Figure 9.3 compares various multimedia data types with transmission bandwidth requirements that provide acceptable delivery of these elements. The figure illustrates the fact that bandwidth requirements increase exponentially as multimedia transmissions move from simple text and graphics to include images, music, video, and visualizations.

Use of the telephone is a good example in this case because it is familiar to all. An average voice transmission across the Plain Old Telephone System (POTS) uses bandwidth ranging from 6 to 44 Kbps. Since basic POTS lines have a bandwidth of 10 Kbps, this means that transmission of music over such a line, while quite feasible, will not result in high-fidelity reception no matter how good the source.

The POTS bandwidth is also quite adequate for transmitting text which requires anywhere from 2 to 10 Kbps. However, if such a line with a slow-speed modem were used for this purpose, the typical page containing about 2 kbytes of text would take about 8 s to transmit. This is comparable to sending a page by fax over the telephone lines, and it is important to remember that such a line cannot be used for any other reason until the transmission is completed.

It's not likely that the networked multimedia applications now arriving will push high-speed LAN technology to the desktop; 10Base-T should be able to handle most video services in the 1- to 2-Mbps range.

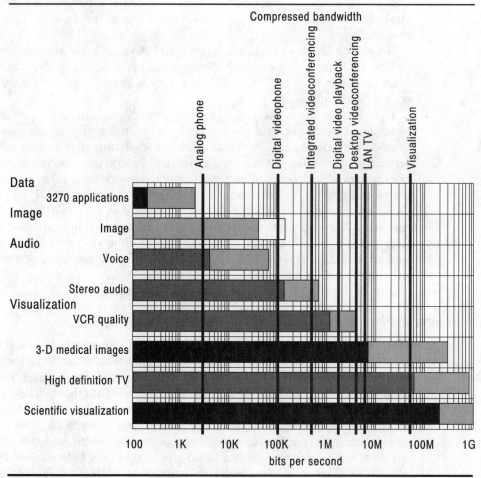

Figure 9.3 Bandwidth requirements of multimedia traffic. (*Source: IBM promotional materials.*)

Theoretically the POTS line could be used to transmit a 24-bit color image of a computer screen with 640 × 480 resolution, but it would take 3 to 4 h to transmit with the lowest-speed modems and many minutes even with transmission lines of much greater bandwidth capacity. Yet in order to reproduce motion comparable to broadcast TV, it is necessary to display 30 such screens every second, which is a task impossible to accomplish without compression and much higher bandwidth transmission facilities.

Audio Latency Issues

In multimedia networking another issue must be addressed from the outset in order to produce acceptable operational results. It concerns human sensitivity to the continuity of audio transmissions and intolerance of even the briefest of interruptions.

Whether the sound is transmitted in real time or accessed in store-and-forward mode, it is absolutely necessary that the audio play continuously. Audio cannot be interrupted even for a fraction of a second without being noticed immediately and almost subconsciously by the human ear. By comparison, in video transmission complete frames can be eliminated either by design or accidentally without affecting the transmission in a noticeable way. With audio transmission this is simply not possible. This issue is of paramount importance when video and audio must be synchronized for simultaneous transmission.

Good speech-grade audio requires bandwidth capacity somewhat better than that of POTS quality ranging between 32 and 66 Kbps. This means that switched 56 or basic-rate ISDN with 128-Kbps bandwidth would be more suitable for such transmissions. Even then, such audio transmission would tie up these lines to the exclusion of other traffic, and these are some of the reasons why compression is also applied to audio transmission.

On the other hand, LAN environment bandwidth does not present an issue in audio transmission. Voice-grade transmissions, even high-fidelity audio of CD quality, which implies 16-bit stereo sampling at 44.1-kHz rate, requires a bandwidth of 176 Kbps, which is well within bandwidth capacities of typical LANs.

A much more significant issue in interactive sound transmission or voicemail is the problem of latency and jitter. Once a person begins a speech or music begins to play, intermittent delays during the session are noticeable and highly annoying to the human ear. It is desirable, therefore, to maintain continuity of sound at all costs in such transmissions. Various studies indicate that the maximum tolerable latency for speech is 600 ms, but experience with satellite communications suggests that a delay of even 250 ms is annoying despite the fact that message coherence is not affected. As a result, it has been accepted that in interactive speech the maximum end-to-end latency that is tolerable and unnoticeable is about 100 ms.

It is important to keep in mind that for transmission paths consisting of LAN-WAN-LAN networks with FDDI backbones, tolerable 100-ms latency is the sum total of all individual latencies in the components of all the networks involved. This situation is illustrated graphically in Fig. 9.4.

Still Images

A simple digitized image screen or frame can vary in size from 10 Kbits to over 500 Kbits depending on the actual image dimensions, pixel resolution, amount

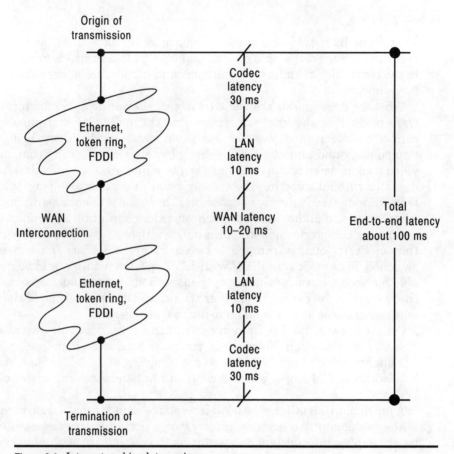

Figure 9.4 Internetworking latency issue.

of detail, and color depth. Average uncompressed images of a full screen contain 512 Kbits, but high-resolution images of photographic quality may require up to 60 Mbits or 120 times more bits. Very high-resolution images are used in various imaging systems in support of medical consultations and diagnoses or when reproducible artwork and illustrations are being transmitted between designers, artists, and publishers.

Color depth plays a very important part in image transmission. An image with 640 × 480 resolution and 24-bit color depth provides up to 16.7 million distinguishable colors and is very desirable in many applications. However, such an image will take three times longer to transmit than the same image with 8-bit color depth which provides only 256 colors. Aside from highly specialized applications, it is obvious that the 8-bit color depth images will be quite adequate in many collaborative multimedia sessions.

Transmission of still images is generally not time-sensitive; therefore, a transmission rate for an uncompressed image with full color depth of 1.2 min on a 128-Kbps channel is quite acceptable by many parties. Even longer transmission rates can be tolerated when image retrieval takes place during off-peak hours or at night. However, some still-image retrieval and transmission may have to accompany real-time multimedia conferencing and collaborative activities. Under such circumstances image compression can be used to significantly reduce the time of transmission. Using the JPEG 10:1 compression ratio, for example, the speed of transmission for the same 24-bit color 640 × 480 image can be reduced from 1.2 min to 7.2 s.

The ultimate application will determine how fast still images should be transmitted within a particular network. As with other multimedia elements, interactive enterprise and virtual corporation environments are unpredictable and their demands on bandwidth cannot be estimated in advance. Nevertheless, in the case of still-image transmissions the solution will always be found in the choice of built-in compression devices within the operating systems. These will provide a selection of compression ratios and transmission speeds at which images can be transmitted within a network. There is a tradeoff, of course, between image quality and compression ratio, but there are also choices of lossy and lossless compression algorithms that are being made available with the latest versions of multimedia software and hardware.

Low-Quality Compressed Video

The videophone is an example of the use of low-quality compression video operating over analog telephone lines and requiring only 100 Kbps of bandwidth in compressed mode. It provides a real-time low-cost interactive point-to-point multimedia videoconferencing device at the push of a button, but the quality of video in these systems is too poor to employ in collaborative multimedia communications.

The quality of the video is affected by frame rate, image size, and compression, which could be lossy or lossless, as discussed in more detail in Chap. 10. In order to save bandwidth, videophone systems deliver a very small image on screens ranging from 1.5 to 6 in in size. In addition, these devices use very low frame rates to create a perception of movement. Where the absolute minimum for continuous movement is about 15 fps (frames per second), videophone systems deliver those very small images at rates in the range of 5 to 10 fps, which results in very jerky and poor-quality video transmission.

Nevertheless, this low-quality compressed video has its place in some applications because it requires a bandwidth ranging from 100 Kbps to 1.5 Mbps and can easily be accommodated by many public carrier facilities. This type of transmission is used when the small image is acceptable for identification purposes or can be used to monitor a series of relatively inactive environments as part of a surveillance operation.

Quality Compressed Video

Quality compressed video transmission, which is time-sensitive, can be divided into two categories according to image quality, which is a function of the compression ratio employed. This in turn depends on the available bandwidth for transmission of such compressed video. The medium-quality compressed video is the result of transmission over facilities with bandwidth capacities up to 1.54 Mbps. The high-quality compressed video is possible with transmission facilities with bandwidth capacities ranging from 6 to 24 Mbps.

Medium-quality compressed video

In this category video is generally delivered at 30 fps, comparable to broadcast TV, but it is usually displayed in $\frac{1}{4}$- or $\frac{1}{2}$-size screen windows or even smaller in some cases. This is the tradeoff to obtain quality video with relatively low bandwidth transmission facilities, which are usually T-1 facilities with a bandwidth of 1.54 Mbps. This is also the medium of choice in business because T-1 lines are commonly employed by MIS departments for linking mainframes and LANs within a corporate environment. As a result, these links are often readily available, but their value may be questionable for sustained videoconferencing if there is a lot of other data traffic with which multimedia transmissions must compete.

As primary-rate ISDN lines become more widely available, these facilities will also provide a bandwidth of 1.54 Mbps comparable to the T-1 facilities. Within telephone companies there are also initiatives to extend T-1 bandwidth capacity to residences for handling video-on-demand services as part of the digital superhighway concept. These are known as *asymmetrical digital subscriber lines (ADSLs)*, and their function is basically to enhance local loop transmission to T-1 levels over 18,000 ft of existing copper lines. It is an attempt of the telephone companies to provide immediate and direct competition to the cable TV companies that already provide coaxial-cable links to many residences.

High-quality compressed video

High-performance workstations with screen resolutions of 1280×1024 and better and operating at 100 million instructions per second (MIPS) or above are the main consumers of high-quality video. This type of workstation is found in engineering and scientific design and research applications, and an increasing number of such units are networked into concurrent engineering or collaborative systems usually running under UNIX operating systems.

High-quality compressed video for these applications requires bandwidths ranging from 6 to 24 Mbps depending on screen resolution and type of computation. This also means that standard LANs providing 10 or 16 Mbps of throughput may not offer sufficient bandwidth capacity to support such applications, particularly when several simultaneous users are involved in a collaborative session. For these transmissions higher-capacity links are used such as T-3 with 46 Mbps or FDDI and other 100-Mbps LANs that are now being developed.

Integrated Videoconferencing

Various forms of videoconferencing exist and have been covered in a previous chapter. The most attractive form which is becoming rapidly popular is multimedia conferencing, sometimes known as *integrated videoconferencing,* that allows users to hold meetings and conferences during which they can simultaneously view and modify documents, charts, graphics, and images. Some of these systems may use two screens dedicated to video and for graphics and data. Multiple users and concurrent meetings can be handled by such systems with compression when transmissions take place using bandwidth in the range of 128 to 348 Kbps. This means LANs and dedicated network interfaces typically switched 56, switched 384, T-1, or ISDN connections.

Such conference systems are designed with the objective of providing integrated rollabout systems that can be connected into carrier networks from various locations. Most systems use specialized codec hardware for compression and provide displays almost always at 15 fps, which is the threshold of motion perception. With higher-transmission-facility bandwidths these systems can display videos at 30 fps and can also communicate with other videoconferencing systems through the use of complex multipoint control units(MCUs).

Desktop Video and LAN TV

Desktop video includes digital video playback from corporate broadcast sources and PC-based videoconferencing systems, which are rapidly gaining in popularity. In 1994, 21,000 desktop videoconferencing systems were shipped. The number of systems in use is expected to reach 7 million during 1998. Codec microchip advances are making it possible to lower prices of these platforms, and already over 30 vendors are offering products in that category. Intel is believed to be spending $100 million on R&D into promotion of desktop videoconferencing chips and products and is seen as a major force in this market. Their objective is to drive down the price of PC videoconferencing by 50 percent every 18 months.

LAN TV is basically video distribution to the desktops to provide current events and enterprisewide information. It includes delivery of TV broadcasts, cable news, and VCR or videodisk library courseware directly to PCs on a LAN. Such transmissions are basically noninteractive and are usually displayed in windows on a PC screen while specialized software provides electronic controls.

This type of video distribution is often accomplished through a separate coaxial-cable facility broadcasting from a corporate headend which obtains inputs from videocameras, VCRs, TV antennas, and satellites. Coupling devices operating on a multiplexing principle are also used for transmitting such video on free bandwidth capacity available on typical LANs.

Bandwidth requirements for PC video systems vary from 128 Kbps to 2.048 Mbps or LAN bandwidth capacities to handle about 6-Mbps transmissions that produce higher-quality results. Most of desktop video falls into the medium-quality video category. These systems are mostly limited to frame rates in the 7- to 15-fps range and relatively small window displays.

Animation

Animation can be defined as a sequence of images which are designed to appear at about 16 fps to create the perception of movement. This frame rate is just above the threshold of jerky motion and creates a satisfactory simulation. However, animation sequences can be developed with any specific frame rate at which objects will move across the screen for creating special effects.

Most animation objects are graphics considerably smaller than the screen, although some animation sequences can be very complex. Designers exercise considerable control over animation sequences which can be arranged to reduce bandwidth requirements without affecting the quality of the presentation. With actual video this option is not available because the image itself is beyond the control of the designer and is a reflection of reality captured by a camera.

Complex high-quality animation sequences require about 20 Mbps, bandwidth and compression is normally used to reduce that requirement. Some animation forms such as those used in complex visualizations, product simulations, and virtual-reality applications require considerably more bandwidth for storage and transmission.

Visualization

Visualization is an important function in engineering and scientific research and product design. It involves 3-D imaging or simulation and animation of complex structures and patterns that are often performed in real time. Virtual reality is another form of visualization which creates cyberspace with which users can interact rather than just observe a 3-D image. Solid modeling and 3-D animation of new product designs is a major application in manufacturing industries using very high-performance workstations and concurrent engineering networking. Within a virtual corporation which is searching for new products and designs to supply a sudden market demand, collaboration with visualization experts will be critical, and this function must be considered in design of such interactive multimedia networks.

Visualization of products involves complex processing, including rendering. This includes many variables such as lighting, shadows, reflections, and movement of objects under changing simulated conditions. Such visualization processes place extraordinary demands on bandwidth capacity and constitute the most demanding multimedia data transmissions.

Visualization in scientific and engineering applications requires bandwidth ranging from 0.64 Mbps for relatively simple applications in chemistry to 800 Mbps in particle physics. On the average scientific and engineering visualizations can range between 100 and 1000 Mbps (1 Gbps). Table 9.3 lists some typical visualization bandwidth requirements in several industries.

TABLE 9.3 Visualization Bandwidth Requirements

Industry application area	Typical bandwidth requirements, Mbps
Engineering imaging	0.1
Chemistry	0.6
Genetics	2.7
Video distribution	6.0
Biological	6.4
Fluid dynamics	16.0
Weather forecasting	40.0
Broadcasting quality video	80.0
Particle physics simulation	800.0

Multimedia Data Compression Issues

In comparison with traditional text and data processing, multimedia objects are typically very large in size and may include images, graphics, animation, video, audio, and visualization components. In the case of some multimedia elements which are time-sensitive, transmission through a network must be continuous. Because of their size, manipulation of such multimedia objects is practically impossible without compression, but this in itself introduces additional delays and complexity into the system.

Interactive enterprises may have to transmit multimedia communications across various and often unpredictable LANs and WANs linking dissimilar systems which nevertheless must appear as uniform and seamless to the end user. This imposes very complex data processing requirements on the networks involved. Such needs are met by high-performance hardware, multiprotocol internetworking products, and international video transmission and manipulation standards. Many of these standards are still in developmental stages, but vendors are developing new products that can handle increasingly complex data types characteristic of multimedia communications.

The Need to Handle Complex Data Types

Multimedia objects impose several functional requirements on data processing infrastructure systems as a result of the necessity to handle very large and complex data types. These present special data representation, data manipulation, data management, and data storage issues. Most hardware and software of legacy and even client-server systems today are not designed to handle massive multimedia data types and must be enhanced with additional hardware and software components or replaced by new devices with built-in multimedia capabilities.

Transmission and storage of such massive and unstructured data, particularly in networked multiuser applications, can create bottlenecks on many existing systems whose bandwidth capacity is often inadequate to handle such volumes which must often be processed in real-time mode.

As a result, there is a need for hardware and software solutions that must include special features for handling the transmission and storage of massive multimedia data without deteriorating the performance of the overall system.

Binary Large Objects

Binary large objects (BLOBs) is the term originally referring to files of images that appeared unusually voluminous relative to conventional textual files. These BLOBs can be B&W, color images from scanners, or specialized image databases, bandwidth-intensive CAD/CAM files and images, multimedia audio and video data, and videoconferencing files. Currently BLOBs are seen as large unstructured data types without a fixed length. They can vary in size from several kilobytes to a gigabyte or more.

BLOB technology has been little understood and seldom used. However, with the explosion of multimedia processing, massive data objects must be stored and manipulated, and vendors of various products are beginning to support BLOBs in their systems.

BLOBs require special consideration because of their size. Conventional storage and databases can store and retrieve only large objects. Any manipulation must be handled within the application itself. For very large BLOBs specialized storage systems such as optical jukeboxes are recommended, but there is a tradeoff in access time that some collaborative multimedia applications may not be able to tolerate. Use of special multimedia servers can alleviate some of the problems caused by BLOBs when transmitting through multiple interfaces and routers. Use of high-bandwidth dedicated circuits such as T-1 with 1.544-Mbps capacity is also helpful. The best solutions, however, for handling BLOBs include compression and careful design of the network itself to make sure that adequate bandwidth is available for processing massive multimedia data as required.

Massive Data Transmission Issues

The massive data processing problem associated with multimedia is best appreciated by analyzing the typical videoscreen because this presents the most complex processing requirements. Digitizing the video itself for multimedia communications is relatively simple. Each pixel of an incoming video signal is analyzed and assigned a digital color value. Using the maximum 24 bits per pixel, a total of 16,700,000 colors can be represented in the palette, which results in a very good-quality image. This means that to represent a whole screen containing $640 \times 480 = 307{,}200$ pixels almost 7.4 Mbits of data is required for a single frame. But in order to create the perception of smooth motion, 30 such frames must be processed with changing pixels every second.

This is equivalent to a torrent of data in the order of 27 MBytes/s without counting any associated digitized audio components. Such volume can increase to about 120 MBytes/s for high-performance workstations with 1280×1024 screen resolutions which are increasingly important in many critical networked business and engineering and scientific applications. The capabilities of existing computer platforms, storage devices, and networks are simply unable to handle such volumes of data because their original designs have been targeted at text and data processing with considerably lower bandwidth demands. As a result, compression is the only practical solution, and some hardware products are already specifically designed to handle massive multimedia data types by including compression microchips on the motherboard of computers and in data communication devices.

Multimedia Continuous Datastreams

Continuous and steady data flow that is required for time-sensitive multimedia transmissions is known as *isochronous information flow*. It is characterized by being continuous and steady and as such permits exchanges to take place in real time such as videoconferencing, videotelephony, and interactive collaborative multimedia conferencing.

Most conventional LAN traffic is asynchronous and is characterized by bursty and packetized data flows. A typical message or datastream is transmitted and stored before it is used. Such a message can contain all types of data, including multimedia components which are not time-dependent. Asynchronous transmissions are taking place along carrier services providing data communications on packet-switching networks.

Time-Dependent Transmissions

Real-time interactive multimedia applications such as various forms of videoconferencing are time-dependent and as such cannot be handled efficiently in the standard bursty mode characteristic of LAN transmissions. Once such a transmission is initiated, it requires a considerable segment of the LAN to be dedicated continuously to that particular user. If these transmissions require about 2-Mbps bandwidth each on the average, this means that once there are five users communicating, a 10-Mbps LAN would be basically unavailable for any other traffic. Time dependence of multimedia traffic comes into play in real-time videoconferencing, but it is also important in store-and-forward video retrieval, and video or TV broadcasting and distribution within a corporation.

These large, continuous, time-dependent multimedia datastreams require that existing LANs be restructured to handle real-time interactive multimedia communications. Enhancements of hardware and software to increase data throughput is one alternative. Replacement of servers and LANs with faster hardware increasing LAN throughput from the current 10 Mbps to 100 Mbps is now a popular solution which is implemented primarily to reduce bottlenecks on LANs and WANs due to increased conventional traffic. In some instances a parallel transmission system can be dedicated to guarantee continuity to video transmission without affecting LAN throughput, but this approach requires integration of video with other data in multimedia applications and may not always be a feasible solution.

Basic Compression Technologies

Interactive multimedia applications cannot exist without compression and decompression schemes in processing multimedia traffic. Existing hardware and networks may have sufficient bandwidth to handle a video transmission, but these capabilities must be considered in conjunction with all other traffic on corporate LANs and WANs that must be accommodated simultaneously. These networks are rapidly being saturated even without multimedia traffic as numbers of users and transactions increase to the point where response time becomes intolerable (Fig. 10.1).

Numerous compression and decompression schemes exist which make multimedia processing a practical reality using hardware, software, or a combination of both. Video data can be compressed in real time as it is being captured and digitized or after it has been stored in the system. The latter implies that sufficient and fast storage capacity is available within the system.

Video compression exploits the fact that the human eye is less sensitive to

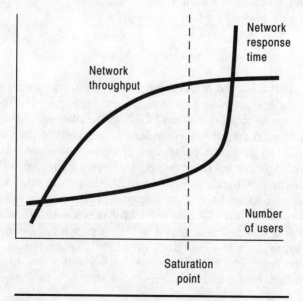

Figure 10.1 Network throughput saturation issue.

color and shade variations than it is to brightness of an image. Compression schemes therefore depend on devising mathematical algorithms that smooth out within digitized video streams all the details that are not normally processed by the human eye. This allows the frame to contain far less data, which in turn reduces the amount of storage, speeds up the transmission of the image, and results in intraframe compression.

There are two forms of compression that can take place. Symmetrical compression takes as long to compress as decompress. Asymmetrical compression takes longer to encode but usually results in higher compression ratios and better quality of output. There are also special compression schemes using fractals which are theoretically capable of compression ratios as high as 10,000:1 but take too long to accomplish to be of value in unpredictable transmissions such as real-time multimedia conferencing.

The basic function of digital compression is identification and elimination of redundant data within and between sequential frames, which results in reduction of data being transmitted. The human eye has a difficult time picking out changes from one frame to another. Eliminating redundancies between frames and only encoding the changes is known as *interframe compression,* which allows much higher compression ratios. The interframe compression compares data in adjacent frames, while *intraframe compression* analyzes similar-appearing areas within a frame. Both methods are used in compression schemes such as JPEG, MPEG, and digital video interactive (DVI).

Compression schemes are differentiated by being *lossy* or *lossless,* which depends on whether all the original data is restored after decompression. Most compression schemes are lossy, but the amount of lost data varies depending on frame rate, final image size, color depth, range of audio frequencies, and quality factors such as sharpness and contrast of the image. This means that many tradeoffs can be made between compression ratios and acceptable quality.

Compression and decompression (codec) can be implemented using specialized DSP microchips, software routines, or both, and these schemes are implemented in various audio and videoboards or embedded within hardware products. Processing time through codecs is critical in calculating the delay or latency in interactive video transmissions. Their existence contributes to the delay during transmission, reception, and in other areas in between when compression processing must take place. This latency is not very critical in stored-and-forward video applications, but studies have shown that cumulatively it should not be allowed to exceed more than 100 ms in a single path for optimal operations.

Discrete cosine transform system

Discrete cosine transform (DCT) is an algorithm which is the basis of many compression techniques now in use (see also Table 10.1 for definition of DCT and other compression schemes). It is a complex mathematical formula which achieves compression by eliminating redundant data in blocks of pixels. Figure 10.2 illustrates the basic compression concept of this type. Although it is similar to the fast Fourier transform, DCT is considered superior in compression because it computes rapidly and yields higher compression levels. There is a

TABLE 10.1   Summary of Major Compression Schemes

Compression scheme	Description of technique
Discrete cosine transform (DCT)	Transform coding technique which is the basis of many compression algorithms such as JPEG, MPEG, H.261, and other popular standards; has practical compression limit of about 230:1
Vector quantization	Basis for DVI compression developed by Intel and most software-based compression algorithms such as QuickTime, Video for Windows, Indeo, Ultimotion, Cinepak, and MotiVE; can be highly asymmetrical
Fractals	Asymmetrical technique with smooth image degradation and theoretical compression ratio at 10,000:1, software products with 2500:1 compression ratios are on the market
Wavelets	AT&T Bell Laboratories technique similar to DCT; uses variable pixel blocks in compression; basis for Captain Crunch and VideoCube codecs
Other schemes	Simple, interpolative, predictive, and statistical coding techniques of compression exist but do not play a major role in multimedia compression since most do not adhere to standards

forward (FDCT) and inverse (IDCT) form of DCT, both of which compute in the same time interval. This is a critical consideration where compression and decompression must be equal in duration and predictable. DCT is the basis for JPEG, MPEG, H.261, and other compression standards.

Compression schemes based on DCT algorithms produce rapidly deteriorating results for compression ratios beyond 200:1 and are believed to be near the practical limit at about 230:1 even though the theoretical limit for DCT is on the order of 800:1. These compression ratios are expected to be in use for the immediate future but it is envisaged that as real-time interactive multiuser multimedia communications become widespread, faster and higher compression schemes will be needed.

Wavelets

Research continues into new compression algorithms such as wavelets and fractals. *Wavelet* compression technology has been developed at AT&T Bell Laboratories and is similar to DCT. It differs in use of pixel block to define areas of detail in an image. Wavelets handle small pixel blocks for compressing fine detail, but bandwidth is saved by use of larger blocks for low-detail areas. This approach is expected to improve compression efficiency, but since wavelets are only a variation of DCT it is debatable whether major compression breakthroughs can be expected from this technology. Wavelet compression is finding application in medical imaging, video editing, and digitized music storage.

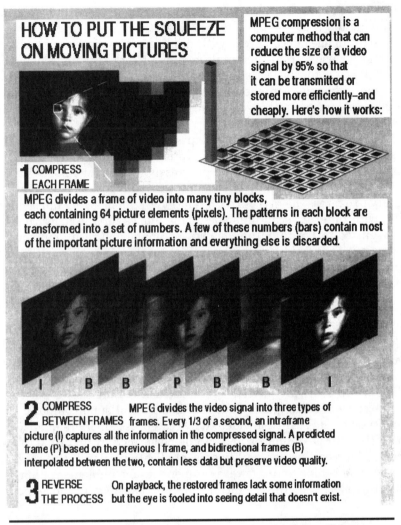

HOW TO PUT THE SQUEEZE ON MOVING PICTURES

MPEG compression is a computer method that can reduce the size of a video signal by 95% so that it can be transmitted or stored more efficiently–and cheaply. Here's how it works:

1 COMPRESS EACH FRAME

MPEG divides a frame of video into many tiny blocks, each containing 64 picture elements (pixels). The patterns in each block are transformed into a set of numbers. A few of these numbers (bars) contain most of the important picture information and everything else is discarded.

I   B   B   P   B   B   I

2 COMPRESS BETWEEN FRAMES   MPEG divides the video signal into three types of frames. Every 1/3 of a second, an intraframe picture (I) captures all the information in the compressed signal. A predicted frame (P) based on the previous I frame, and bidirectional frames (B) interpolated between the two, contain less data but preserve video quality.

3 REVERSE THE PROCESS   On playback, the restored frames lack some information but the eye is fooled into seeing detail that doesn't exist.

Figure 10.2   Data compression concepts. (*Source: Reprinted from February 14, 1994, issue of* Business Week *by special permission, copyright © 1994 by McGraw-Hill, Inc.*)

Fractals

Fractal is short for *fractional dimensional,* which is a mathematical term describing a fractional element of an image after repeatedly applying a specific compression algorithm. Fractals are an asymmetrical compression technique and result in smooth-image degradation because they employ repeatedly relatively simple algorithms in the compression process. They have a theoretical compression ratio of 10,000:1 and commercial software products using fractal compression ratios of up to ~2500:1 have been on the market. As with all compression schemes, an acceptable compression ratio remains the function of image quality that the end user will tolerate.

Fractal images decompress with minimal delay and thus are of particular interest for use in real-time multimedia network transmissions where excessive latency is always a threatening problem. Fractals may also prove useful for identifying the contents of complex images in multimedia databases as archives of images and videos resulting from increasing multimedia traffic continue to accumulate.

Fractals search patterns and shapes within an image, determining mathematical functions that represent transformations of the image. They are applied repeatedly and generate progressively smaller images until they converge on a unique fractal. The compression advantage of fractals lies in the ability to express the process with relatively simple algorithms applied repeatedly with differing scaling factors. This allows for the algorithm itself to be transmitted rather than the compressed image and results in very significant reduction of bandwidth requirements. Fractal compression, although in use, is still in developmental stages and research is under way to improve their methods for defining fractal algorithms in specific images.

Digital video interactive

Digital video interactive (DVI) is a programmable compression technology based on vector quantization originally conceived at the RCA Sarnoff Laboratories and developed later by Intel and IBM. It is an asymmetrical technique which offers production-level video (PLV) compression with a 120:1 ratio and excellent image quality but requires special supercomputer facilities for compression. Its real-time video (RTV) compression offers quality comparable to Motion-JPEG standards and can be performed in real time allowing interactive video editing for multimedia applications. DVI compression requires special microchips and software components for implementation, but it is no longer supported by Intel, which is now emphasizing its Indeo technology.

Software compression schemes

In 1991 Apple and SuperMac Technology collaborated to develop QuickTime, a software codec that offers a choice of compression techniques and can compress digitized video. A similar software compression codec was later announced by Microsoft as Video for Windows, while Intel followed with Indeo and IBM with Ultimotion. Most software codecs are based on the *vector quantization technique,* which basically follows the MPEG compression standard but does not employ DCT algorithms. This method identifies blocks of similar images and replaces them by an average picture.

Although known as *software codecs,* many require video digitization hardware to operate. Most can achieve compression which will provide video frame rates of 30 fps for small (160 × 120) windows with 16-bit color depth. Larger video size requires a reduction in frame rate which can be scaled down to the marginally acceptable 15 fps.

Other compression schemes

Research and development on image and video compression has been going on for over a quarter of a century. It resulted in numerous techniques that are not compatible with each other. DCT is only one such method of several transfer coding techniques that has been widely used as a basis for video compression, but because of its practical limitations, new methods are constantly investigated. Other compression techniques in existence are classified as simple, interpolative, predictive, and statistical but do not play a role in multimedia communications, which rely heavily on standards that are mandatory for widespread use of multimedia technology. This does not mean that some of those techniques will not be adapted in the future to interactive multimedia communications if they offer an advantage in some of their operating versions.

A Question of Compression Standards

A major issue in multimedia audio and video transmission is interoperability across the various platforms and data communications equipment. This problem can be particularly acute when several interactive enterprises decide to engage in a virtual corporation type of activity which requires interactive multimedia communications between new and unknown network systems. With a large number of compression schemes implemented by many vendors in the hardware and software products, standardization of the process is of paramount importance. Designers of multimedia networks must pay special attention to the components of their systems and make absolutely certain that all versions of even the same-sounding compression schemes included in their equipment are in fact compatible with each other. The situation is improving from the early days of multimedia computing, and many codecs in fact contain a choice of the most common compression schemes which are automatically brought into play when the system detects specific requirements for their implementation.

Several industry groups and international committees have been developing standards for image, audio, and video compression over the last few years, and this work continues. JPEG, MPEG, ISO, and CCITT are the major bodies involved, and networked multimedia communications developers should keep track of the latest standards as these are being developed and adopted.

Joint Photographic Experts Group (JPEG) standard

This standard was developed specifically for digitizing, compressing, and decompressing still B&W or color images from cameras, videocameras, scanners, fax machines, and other office automation equipment. JPEG compression ratios range from 10:1 to 80:1, but JPEG compression presents a continuous tradeoff between quality, speed of delivery, and software capabilities. JPEG compression is based on the DCT algorithm but is incompatible with the standards for still-frame graphics developed by CCITT, although it will be included in the H.261 standard.

Using JPEG compression at 30 images per second can substitute for motion video compression. This procedure, known as *Motion-JPEG,* has become a popular implementation of digital video compression before MPEG compression was fully developed. The drawback of Motion-JPEG lies in the fact that as an accelerated still-image compression algorithm, it does not handle the associated audio components. Nevertheless, numerous Motion-JPEG boards have been introduced into the marketplace by vendors based on the 550 JPEG chip developed by C-Cube Microsystems. Some software codecs also implement JPEG compression.

Motion Picture Expert Group (MPEG) standard

MPEG-1 is now an established standard for digital video compression for use with interactive multimedia applications, and MPEG-2 is designed for broadcast TV and video-on-demand transmissions. MPEG is based on intraframe and interframe motion estimation techniques. The MPEG algorithm analyzes data in key frames known as *I frames* or *intrapictures.* Only limited compression is obtainable from such frames, but *P frames* in between eliminate redundancies with reference to I frames and build a foundation for future P frames. This means that compression is achieved by recording only changes from frame to frame.

MPEG is also based on the DCT algorithm and was originally applicable to video and audio with no more than 1.2-Mbps bandwidth. MPEG-2 can support broadcast-quality video and handles data rates of up to 15 Mbps, which includes the typical LAN bandwidth spectrum. The best compression results are obtained with MPEG hardware implementations, but MPEG software for video playback in 320×240 and 240×120 resolutions also exist. In the spring of 1993 the first real-time MPEG capture and compression 8-bit video card was introduced by Xing Technology and Philips Semiconductor Company.

H.261 International videocoding standard

This standard pertains specifically to videoconferencing communications throughout the world in an attempt to facilitate interoperability between incompatible systems. Until its announcement in 1990, videoconferencing vendors used proprietary codecs incompatible with each other optimized for transmissions on private lines in the 64- to 384-Kbps bandwidth range.

The H.261 standard, also known as the *p*64-Kbps* standard, where $p = 1,2,...,30$, outlines the process that all codecs must follow in compressing digital video for transmission. It specifies pixel blocks and groups of blocks for compression, image resolution, video data formatting, and coding tables required. It also specifies how codecs of any vendor can communicate with each other at frame rates of up to 30 fps, which is comparable to broadcast TV quality.

H.261 also specifies standard video display formats with *common intermediate format* (CIF) at frame resolution of 352 × 288 pixels. It also recognizes the *quarter common intermediate format* (QCIF), which is 176 × 144 pixels and can operate at 15 fps, providing an acceptable video reception. These standards are designed to provide reasonable videoconferencing transmissions over 112- and 128-Kbps lines. Figure 10.3 illustrates the relation of H.261 standard frame sizes to those of typical PCs and workstations.

G.728 audio compression standard

Because human beings are much more sensitive to sound variations than to video quality, audio digitization and compression techniques are already well established. Quality of digital sound varies according to sampling frequencies and quantization, which means the number of bits used to define a sample. For ordinary speech the audio signal is sampled 8000 times per second at 8-kHz sampling rate and 8-bit coding is used to represent 256 different amplitude values of the sound. This form of encoding will require 64-Kbit bandwidth for transmission.

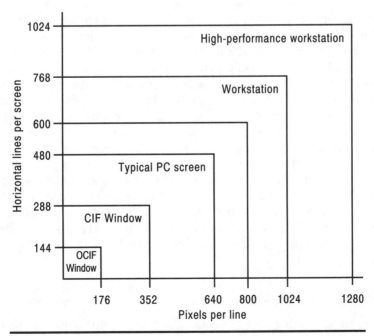

Figure 10.3 H.261 Standard videodisplay formats. (*Source: 21st Century Research*)

AM and FM (amplitude and frequency modulation) broadcast quality sound may require sampling rates of 18.9 kHz with 8-bit coding but high-quality music requires 44.1-kHz sampling rates and 16-bit coding, which significantly increases the bandwidth required for transmissions. At this level one minute of stereo with 16-bit coding and 44.1-kHz audio requires 10 Mbytes of storage. Audio must remain a continuous signal and is one-dimensional, which means that compression ratios are considerably lower than those for video, mostly in the 4:1 range. Efforts are under way to develop 8:1 ratios for audio which would allow high-fidelity sound transmissions at 64 Kbps. The latest audio standard for international transmissions, which was ratified in April 1992, calls for audio compression at 16 Kbps.

Codecs

The basic engine that drives multimedia transmission, particularly in the videoconferencing industry, is the codec (coder-decoder), which is the video counterpart of a modem. Codecs take analog audio and video signals from video cameras and other sources, digitize them, and compress the digital values for transmission at lower bandwidth. At the receiving end codecs decompress the digitized signals, convert them back to analog, and transmit them to audio and video output devices. Current codecs operate with 384-Kbps lines, which represent a compression rate of 230:1, and some codecs also operate with 112-Kbps circuits, which means compression rates of 800:1.

The highest-quality codecs offer more than 175,000 pixels-per-frame resolution, which is a measure of video quality that is so critical in multimedia video applications. This is close to top-quality-broadcast TV video quality, which ranges from 190,000 to 200,000 pixels per frame. The quality of video images also depends on how often the screen is refreshed; the standard TV video is 30 fps. Codecs reduce the frame rate to 15, 10, or even 5 fps depending on the amount of motion, but most codec vendors claim a maximum potential of 30 fps. In general users cannot perceive a difference in video quality until the frame rate falls to about 18 to 16 fps. At rates typically offered over the 112- and 128-Kbps circuits, the frame rates are in the order of 10 to 15 fps, and some deterioration of video quality is noticeable when fast and continuous motion is involved. Pre- and postprocessing of the data and motion compensation techniques further reduce the amount of data handled by codecs, and these methods are key in providing acceptable video quality in 112- and 128-Kbps facilities, which are also available at lower tariffs than those with higher bandwidth.

Major Suppliers of Data Compression Products

There are several categories of vendors offering compression products. These include specialized compression microchips, compression software solutions, compression boards based on codecs integrated with audio and video digitizers, and traditional videoconferencing companies with proprietary codec systems.

The basic compression microchip manufacturers include AT&T, C-Cube Microsystems, Compression Labs, Duck Corporation, Intel, and LSI Logic.

Vendors who incorporate basic compression microchips in their products such as audio and video boards include Apple, Creative Labs, Digital F/X, Horizons Technology, IBM, Intel, Matrox, Media Vision, New Media Graphics, New Video, Optibase, Radius, RasterOps, Storm Technology, SuperMac Technology, Truevision, and Videologic. However, it must be understood that many of these boards are specifically designed for delivery of multimedia applications and primarily perform only the decompression function in multimedia transmissions.

Software compression solutions using various schemes are available from companies like Apple, Aware, Image-In, Intel, Iterated Systems, IBM, LEAD Technologies, Microsoft, Optibase, Optivision, Telephoto Communications, Video & Image Compression Corporation, and Xing Technology.

Traditional videoconferencing firms which pioneered codec development include Compression Labs, PictureTel, and VTEL are also developers of compression devices, but in many cases those companies collaborated with specialized semiconductor manufacturers to design proprietary codecs for use in the videoconferencing systems.

Multimedia Network
Transmission Standards

The term *interactive multimedia communications* implies enterprisewide, regional, national, and even global telecollaboration between individuals and groups of workers. As a result, multimedia objects will not only vary vastly in size and nature but also originate from widely scattered locations and a great variety of sources. The need to handle massive amounts of information in itself presents only a throughput problem, which is being addressed by faster hardware and compression and higher-bandwidth transmission lines.

Challenging as this is, interoperability between groups, enterprises, and countries may present a set of even more difficult issues to resolve. These include synchronization of multimedia objects, real-time handling of collaborative sessions, and general inadequacy of communications service facilities to handle time-dependent video transmissions.

What this means is that international communications networks must develop and provide high-capacity digital data links, variable bandwidth capabilities, and means to synchronize different categories of multimedia traffic. In order to streamline existing and new carrier services to provide such facilities, an extensive set of standards governing multimedia data transfer between all collaborating parties must be in place.

Standards Organizations Affecting Multimedia

Industrial, national, and international organizations are involved in proposing, studying, recommending, and ratifying standards that affect multimedia traffic. These are all de jure standards sanctioned by governments and international bodies, but there are also de facto standards imposed by one or more manufacturers designed to capture and maintain a leading market share in a particular niche. Besides the JPEG and MPEG compression standards dis-

TABLE 11.1   Standards Organizations Affecting Multimedia

Organization	Description of major activities and objectives
ANSI (American National Standards Institute)	Private nonprofit domestic organization; oversees standards setting groups in USA; inputs from EIA, IEEE, and NBS
IMA (Interactive Multimedia Association)	Professional trade association; provides open forum between developers and vendors of multimedia products
ISO (International Standards Organization)	International body since 1947 in Switzerland; fosters exchange of science and technology information; jointly with CCITT, set up JPEG and MPEG groups to design compression standards
ITU (International Telecommunications Union)	International body operating since 1932 under UN; its CCITT committee created the H.261 (p*64) videoconferencing standard and the H.320 family of audiovisual telephony standards; collaborating with ISO in JPEG and MPEG

cussed in the previous chapter, multimedia data transfer standards are influenced and adopted by organizations such as ANSI, IMA, ISO, and ITU. Table 11.1 summarizes the activities of the main standards bodies involved.

American National Standards Institute (ANSI)

ANSI is a private nonprofit domestic organization which oversees the efforts of all public and private standards setting groups in USA. In matters dealing with multimedia ANSI is influenced primarily by the Electronic Industries Association (EIA), the Institute of Electrical and Electronic Engineers (IEEE), and the National Bureau of Standards (NBS). In addition, there are about 300 standards committees that have been organized outside ANSI to carry out specific standards tasks. Two of those that have the most impact on multimedia technology are the Accredited Standards Committees (ASC) on telecommunications and on computers and information processing. ANSI also has a mandate from the U.S. State Department to represent the United States in the international arena.

Interactive Multimedia Association (IMA)

This organization was launched in 1988 as a professional trade association of individuals and organizations involved in production and use of interactive multimedia technologies. IMA is a standards-setting body supported by most multimedia developers and serves as an open forum for vendors and users. Its primary goal is to identify various multimedia system protocols and specifications and to obtain a consensus among developers. IMA promotes the development of interoperability between different multimedia systems.

International Standards Organization (ISO)

Based in Switzerland, ISO in its present form has been operating since 1947, representing 90 different nations. Its objective is to facilitate the exchange of scientific and technical information to assist in development of international trade. The most pertinent group concerned with multimedia is the Technical Committee 97 (TC97), which deals with computer and information processing standards. Because of the nature of multimedia, other committees dealing with standards for broadcasting and telephony are also involved. ISO collaborates with CCITT in the JPEG and MPEG groups for compression standards.

International Telecommunications Union (ITU)

This is the standards-setting body formed in 1932 which came under the control of the United Nations in 1947 and is based in Geneva, Switzerland. Two of its committees have a direct impact on multimedia data transfer. These are the Consultative Committee on International Radio (CCIR) and the Consultative Committee on International Telegraphy and Telecommunications (CCITT; now renamed ITU-T). This organization is responsible for the creation of the H.261 or p*64 standard specifically designed for videoconferencing in 1990 and the more recent H.320 family of standards pertaining to audio and video telecommunications between equipment from various manufacturers.

The Key H.261 Videotransmission Standard

The key standard for videoconferencing and multimedia interoperability is H.261, which defines several procedures for codecs to communicate at frame rates of up to 30 fps over lines consisting of multiples of 64-Kbps bandwidth facilities.

CCITT adopted the H.261 standard, also known as p*64, specifically designed to define transmissions of video images over digital networks at data rates ranging from 64 to 2048 Kbps. Until then videoconferencing vendors used proprietary formats with various compression techniques that were incompatible with each other and optimized for videoconferencing transmissions on private lines. H.261 includes a compression standard that is based on the DCT technique and is similar in nature to MPEG but allows much faster transmission speeds. H.261 sacrifices video quality and image size in order to provide videoconferencing interoperability in QCIF and CIF formats on lower-bandwidth lines of 112- and 128-Kbps capacity.

Originally this standard was designed for control of point-to-point videoconferencing, but the use of this technology now overlaps multimedia applications. With collaborative multiparty conferencing, new functions come into play that require additional transmission standards. As a result, the H.261 standard is being expanded to cover a number of new critical functions. These include multipoint bridging and internetworking security, as well as multimedia applications and PC connectivity with LANs and WANs.

H.261 Still-Frame Graphics Standard

A standard for coding still-frame images is one area of multimedia communications that has not been defined but is seen as an option for H.261. Applications that would benefit from such a standard include concurrent engineering and publishing applications. Such a standard would allow different codecs to exchange information about complex engineering drawings, charts, or maps that are inspected on the screen during a conference. Still graphic images, however, usually require higher resolution to show details with better precision than do captured video images. Current codecs capture the still image of such a drawing, compress it, and transmit separately from the real-time video. The optional H.261 standard will specify a method for acquiring additional bandwidth from the video channel to improve the resolution of such transmissions. Such a feature is critical during multiparty conferences when all participants can take part in annotating and changing details on such images.

H.231 and H.243 Multipoint Videoconferencing Standards

Two critical new standards under development address multipoint bridging in networks. Multipoint conferencing is the logical extension, enhancing relatively simple point-to-point videoconferencing to handle several parties simultaneously.

The H.231 standard is designed to control linkups between three or more dissimilar codecs with multipoint control units (MCUs). The H.243 standard controls procedures between H.231 MCUs. The key to H.243 standard are *bit-rate allocation signal* (BAS) codes that allow codecs to switch seamlessly between point-to-point and multipoint conferencing.

Multipoint conferencing standards should also include the rules governing acquisition of new participants during a multipoint conference and departures of participants while the videoconference is still under way. Control of side conferences between some participants while the overall videoconference is under way is also a significant issue in such environments that must be addressed with a set of standards.

H.233 Multimedia Transmission Encrypting Standard

The H.233 standard specification defines a method for use of a unique encryption key between authorized videoconferencing parties before any collaborative activity can take place. It is not yet fully understood what information and data security issues are involved in multiparty multimedia conferencing. These, along with multiuser access to corporate databases, sensitive marketing intelligence, and personnel records, must be protected for confidentiality at various levels of multipoint conferencing.

Such issues are very critical in health care multimedia communications and must be resolved before these systems become widely used. There are many images and videos in such medical environments that contain highly sensitive information about patients, such as their health, diseases, medical therapies, financial status, and associated physical and psychic conditions. The encryption standard may turn out to be one of the most important developments that must take place before virtual corporate activities can get under way.

The H.320 Umbrella CCITT (now ITU-T) Compliance Standard

The H.320 is the umbrella CCITT compliance standard which specifies all the technical requirements for narrowband transmission of audiovisual systems. H.320 implies compliance with the basic H.261 video coding standard, a number of auxiliary standards, and a new audio compression standard. These component standards include H.221, which deals with framing information; the H.230, for handling control and indication signals; and the H.242, which governs transmission setups and disconnects. The new G.728 standard, which deals with audio compression at 16 Kbps, is also a part of this family and was discussed in more detail in the previous chapter. Table 11.2 summarizes the entire H.320 family of standards adopted and in development.

All the H-series standards have been adopted in 1990, while the G.728 was finalized only in 1992. It is clear that this family of standards is still growing and the multimedia protocol and MHEG standards may well become a part of H.320 when these are adopted. The H.320 complete suite of standards is now available on a single VCP microchip from Integrated Information Technology

TABLE 11.2 Summary of Video Transmission Standards

Standard	Description	Status
MPEG-1	Motion video for multimedia	Adopted
MPEG-2	Broadcast version of MPEG-1	Adopted
JPEG	Still-frame graphics for multimedia	Adopted
Motion-JPEG	JPEG moving video animation	Widely used
H.261	Video coding also known as p*64	Adopted 1990
H.261 Option	Still-frame graphics option	Proposed
H.320	Overall requirements for N-ISDN	Adopted 1990
H.221	Framing information	Adopted 1990
H.230	Control and indication signals	Adopted 1990
H.242	Terminal capabilities exchange	Adopted 1990
H.233	Encryption and security aspects	In development
H.231/243	Multipoint videoconferencing	In development
No number	Nonvideo data transmission	Proposed
G.711	64-Kbps pulse code modulation audio	Adopted 1984
G.722	48/56/64-Kbps adaptive differential PCM	Adopted 1986
G.728	16-Kbps audio transmission	Adopted 1992

which also includes JPEG, MPEG, and other proprietary algorithms. AT&T also developed the AVP set of microchips which perform the same function.

Multilayer Protocol Standard

An important addition to the H.320 suite of standards is the *multimedia protocol standard,* which is under development at ITU under the code name T.120. It is going to cover standards for PC-based multimedia networking and conferencing. This standard deals with multilayer protocols that are critical in collaborative activities. This new standard will cover multiparty videoconferencing including transmission of word processing documents, spreadsheets, databases, and other nonvideo components.

Multimedia and Hypermedia Information Coding Expert Group (MHEG)

This is one of the most recent standards being developed by ITU and ISO specifically addressing the synchronization problems that can occur when compound multimedia documents are transmitted over WANs. Such problems are usually caused by differences in size of files being transmitted. Image files are perfect examples of BLOBs because they are normally many times larger than text files and take much longer to transmit. It is these differences in size that often cause transmission problems. MHEG is the proposed new standard which will address the method of linking different components of a multimedia or hypermedia BLOB or document, particularly in instances where one element triggers the presentation of another.

Multimedia LAN Alternatives

Although existing LAN standards offer relatively high bandwidth capabilities, they are not suitable for handling time-sensitive multimedia transmissions required to support multipoint collaborative operations.

In 1994 about 21 users existed on an average LAN, which automatically limits the available bandwidth per user to less than 0.5 Mbps, whereas even a compressed video transmission requires about 1.5 Mbps and continuous availability of the channel. The average number of users per LAN continues to grow, and with increasing use of graphics and images, the transmission capabilities of existing LANs are being stretched to the limit. Organizations are already considering higher-capacity LANs to handle their existing traffic, which consists mostly of text and data.

The most popular networks such as Ethernet and token ring are designed to process data packets and operate in contention modes. This is not conducive to the effective transmission of continuous streams of time-dependent multimedia data characteristic of most videoconferencing and collaborative applications.

In short, the existing LAN technologies are not adequate to support enterprisewide interactive multimedia collaborative activities that are the basis for the interactive enterprise and participation in virtual corporation types of activities. This is not to say that individual LANs cannot handle interactive multimedia communications, but in real-time applications in competition with all the conventional text and data traffic, this does not appear as a practical possibility.

However, existing LANs can be modified to handle multimedia transmissions effectively including interactive multiparty conferences through a number of alternatives. These include microsegmentation, isochronous Ethernet, high-speed LANs, intelligent-hub solutions, multimedia capable PBXs, or ATM switching technology. In all cases additional costs are incurred for specialized hubs and interfaces.

Multimedia LAN Requirements

Numerous LANs that are in operation today offer bandwidths ranging from 4 to 16 Mbps and can handle digital high-speed communications within workgroups and departments. Such LANs generally are limited to communications within a building or nearby facilities no farther than a few miles away. LANs are also designed to handle intermittent, bursty traffic of small data packets, rather than continuous datastreams that are required by multimedia video and audio transmissions.

LAN users are now accustomed to fast and efficient transfer of information over those networks, although most of it consists of textual and data objects in store-and-forward modes. There is already a growing demand for higher throughput driven by an increasing number of users and the spreading use of graphics and imaging. In a few years storage and retrieval of multimedia objects for interactive applications is expected to skyrocket. Peak-bandwidth-demand requirements for such multimedia traffic will be hundreds or even thousands of times greater than that for conventional text and data interactions. Users will demand, nevertheless, that such massive multimedia traffic is processed as rapidly as conventional LAN interactions. This means that existing networks must be enhanced to handle such traffic or new infrastructures must be developed to replace inadequate network architectures.

Hardware manufacturers have made it relatively simple and inexpensive to add multimedia capabilities to standalone PCs and workstations. It is now inevitable that end users equipped with multimedia-capable platforms will sooner or later demand multimedia out on the network. That's when they realize that audio and video transmissions are time-sensitive and cannot be processed efficiently in the existing bursty LAN operating modes. Video or TV distribution, whether real-time or relayed, and store-and-forward video transmissions must also be continuous while multimedia conferencing must be real-time and continuous simultaneously. As a result, there is a need to restructure existing architectures to handle real-time interactive multimedia communications and to make sure such transmissions can be implemented without any intolerable delays or latencies.

On the other hand, existing LANs represent considerable investments by their users. A major challenge for network administrators in the immediate future is to develop LAN and internetworking strategies that permit gradual enhancement of the infrastructures to handle multimedia communications capable of supporting interactive enterprises and virtual corporate operations when necessary. This approach will provide the maximum investment protection in existing LAN and WAN systems.

Multimedia Potential of LAN Topologies

The *topology* of a particular LAN is defined as the type of connectivity between workstations or PCs that are networked. Several LAN topologies exist including bus, star, ring, tree, and mesh formats; the bus and the ring are the most common. Figure 12.1 illustrates the basic configurations of these topologies.

Uniform protocols and interfaces allow easy connection of dissimilar networks whatever the LAN topology. Multiple LANs are interconnected into corporate campus, metropolitan (MAN) or wide area networks (WANs) which are linked with high-bandwidth fiberoptic backbones such as FDDI or ATM. WANs, in turn, are linked through private or public digital services with other WANs or networks throughout the world if necessary.

The nature of end-user connectivity to LANs varies depending on the topology, and not all topologies are suitable for use with interactive multimedia traffic. However, with additional expenditure, special high-bandwidth interfaces, routers, or intelligent hubs, any topology can be reconfigured to handle interactive multimedia traffic. Since expansion of bandwidth capacities of LANs is taking place, anyway, the issue becomes one of making sure that future plans take into account the nature and volume of multimedia traffic that may be expected.

The bus topology

The *bus* format consists of a common shared bus line with a bandwidth of 10 Mbps. Individual stations are directly connected to the bus and signals travel in both directions on the bus from the point of origin until being dissipated at terminations. Any station transmission is detected by all other stations, so a bus is like a broadcast operation. If a station tries to transmit uncompressed video requiring about 30 Mbps of bandwidth, it will close down the LAN. If only a few stations transmit compressed multimedia video streams requiring up to 2 Mbps the LAN would not be able to handle this volume of data in such a configuration.

The ring topology

In the *ring* topology each station is connected to one preceding it and one ahead of it. This results in a continuous circuit through which signals travel in a single direction. In this arrangement packets sent from an upstream station are repeated by the computer to its downstream neighbor. These data packets travel in a single direction until they are removed by the original sender. The ring topology requires bypass switches at each station to continue operating when one or more of the stations are not active. IBM token ring LANs, which can provide 4 or 16 Mbps of bandwidth, are the best example of this format. In practice this means that five or six simultaneous compressed video transmissions can take place, but this can happen only at the exclusion of any other traffic on that LAN.

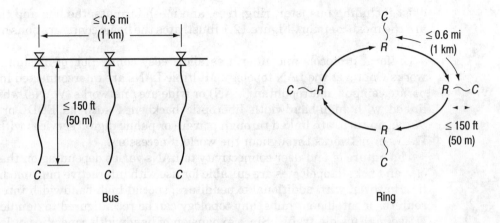

Bus   Ring

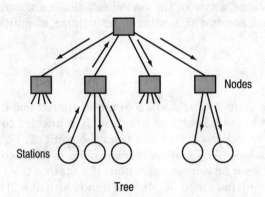

Tree

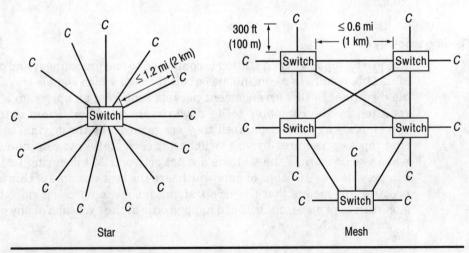

Star   Mesh

Figure 12.1 Basic LAN topologies.

The tree topology

The *tree* topology is characterized by a hierarchical pattern with multiple branches leading to individual stations. In this configuration a station transmits only to a higher-level network node, which passes the message to another station or a still-higher-level network node. In such a topology the network is only partially loaded when a transmission takes place. The local telephone system is a good example of such a configuration. This format makes the tree topology of potential interest for multimedia communications. In fact the LAN segmentation approach to multimedia networking takes advantage of this configuration to reconfigure a LAN into a tree topology.

The star topology

The *star* topology can be considered a special case of a single-level tree topology in which each station is connected to a central node or hub. This hub directs the flow of multimedia traffic, routing transmissions to other stations. In that sense it becomes a broadcasting system and is logically similar to the bus topology. All the problems associated with multimedia traffic on a bus topology will exist unless the central node is an intelligent hub designed to discriminate between various transmissions. Some star topologies have individual stations connected to each other to form a ring. This implies all the problems associated with that topology. In some instances a tree topology may be seen as an outer layer of a microsegmented star.

The mesh topology

The *mesh* topology is among the most complex configurations. In this format stations may have direct links with most other stations. The mesh topology can be compared to ring topology with partial tree or star topology patterns. As such, it offers limited potential for direct multimedia traffic handling because many links must be upgraded to handle multimedia. However, as with all the other topologies, it can be partially or totally reconfigured with intelligent hubs at an additional cost.

Enhancing Existing LANs for Multimedia

All existing LANs have the theoretical bandwidth to handle multimedia transmissions, but they do not offer practical solutions in their present form. In order to develop enterprisewide video networks capable of handling collaborative multimedia activities, considerable restructuring of existing LANs is necessary.

The interactive enterprise requires a networking infrastructure that can bring together a number of multimedia-related technologies. These include new multimedia platforms at the desktop and in the field, specialized videoservers for transporting analog and digital video to users, conventional

Enterprise video networks will bring together a wide range of technologies, mixing conventional videoconferencing units and MCUs along with desktop multimedia and a new class of videoservers.

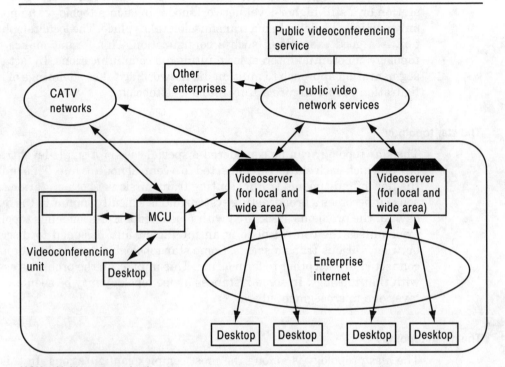

MCU = Multiport control unit

Figure 12.2 Elements of a multimedia network. (*Source: Reprinted from* Data Communications, *February 1993, p. 62, copyright by McGraw-Hill, Inc., all rights reserved.*)

videoconferencing systems with their associated MCUs, and massive new storage facilities. Figure 12.2 illustrates the elements of such a multimedia network.

If an interactive enterprise wants to collaborate with other such units outside its parent environment or combine into a virtual corporation activity even temporarily, new networking infrastructures must come into being. These will have to include a wide range of interfaces for linkage readiness with other enterprises and circuit-switching public videoconferencing services, as well as cable TV and other video broadcasting transmissions. The overall objective in enhancing existing LANs must be as follows:

- Engage in multipoint multimedia conferencing across a wide area in a seamless manner.

- Share multimedia files with other users within and without the enterprise during collaborative sessions.

■ Distribute cable TV or other broadcast and specialized video from internal and external sources.

■ Provide specialized massive multimedia storage facilities for corporate and individual usage.

The key to success in providing for interactive multimedia services as listed above requires redesign of the LANs to handle real-time video traffic at random without intolerable delays and use of specialized videoservers coupled with massive amounts of hierarchical storage probably in the category of redundant arrays of inexpensive disks (RAID) and optical jukeboxes.

As far as enhancement of the existing LANs is concerned to make them multimedia-capable and meet such objectives, a number of solutions are available. These include microsegmentation of the LAN, isochronous Ethernet, and high-speed LANs such as Fast Ethernet, CDDI, FDDI, HPPI, and ATM (Table 12.1).

The LAN Segmentation Solution

One of the simplest solutions for extending networked multimedia applications is microsegmentation, which can provide uncontested local bandwidth at the desktop. *Microsegmentation* is an architectural strategy that limits the number of users on a LAN segment in relation to each user's bandwidth requirements. Under such an arrangement an entire LAN segment could be claimed by a single user creating, in effect, a desktop LAN. The advantage of this solution lies in the fact that it is normally implemented to relieve LAN congestion regardless of whether multimedia transmissions are to be accommodated.

TABLE 12.1 Summary of Multimedia LAN Alternatives

Multimedia LAN	Description of concept and comments
Microsegmentation	Architectural solution which limits the number of users on a LAN segment in relation to bandwidth requirements; also used to relieve congestion on LANs
Isochronous LAN	Enhancement of existing LAN by addition of a 6-Mbps isochronous channel dedicated for voice and video
Fast Ethernet	Provides 100-Mbps bandwidth in two variants: (1) 100Base-X compatible with existing Ethernet mode; (2) 100Base-VG uses dedicated voice-grade lines for isochronous transmission
FDDI	Offers 100-Mbps bandwidth in four variants: (1) asynchronous FDDI—includes no prioritization of traffic; (2) synchronous FDDI—can handle time-sensitive multimedia traffic; (3) FDDI II option—a new multiplexing architecture; (4) CDDI—implementation of FDDI on copper wires with limited range
PBX	Circuit-switching and hybrid PBX models can be upgraded to carry video transmissions by dedication of channels
ATM	Asynchronous transfer mode is considered ideal solution for LANs and WANs but requires costly interfaces, hubs, and switches

In order to implement microsegmented LANs, existing shared network topologies must be restructured into a two-tiered star topology with switching both between individual LAN segments and within the backbone connections and hubs. Each LAN segment is linked directly to a segment-switching hub which is associated with a local videoserver, and each hub is linked directly to a main hub which provides hub-to-hub connectivity and access to local and wide area networks. Figure 12.3 illustrates the concept of microsegmented LANs for handling multimedia traffic.

As microsegmentation is used to create desktop LANs, today's shared network topology will evolve into a two-tiered star, with switching both between LAN segments and within the backbone.

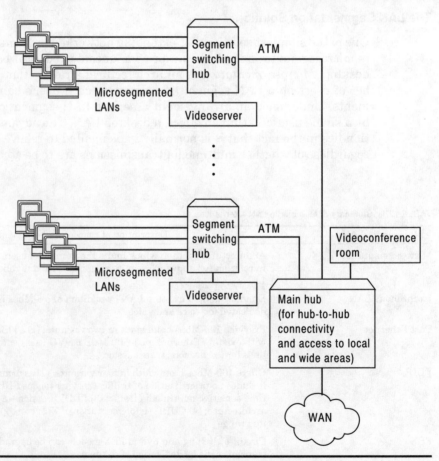

Figure 12.3 The microsegmented LAN. (*Source: Reprinted from* Data Communications, *February 1993, p. 66, copyright by McGraw-Hill, Inc., all rights reserved.*)

Segment-switching hubs create an aggregate network bandwidth equal to the cable bandwidth multiplied by the number of segments. As a result, a 10-Mbps Ethernet LAN with 10 segments will appear to provide the equivalent bandwidth of 100 Mbps. A similar effect can be achieved with Ethernet switches that are now on the market. These are devices that provide 10-Mbps bandwidth per port and have the potential to multiply Ethernet bandwidth almost indefinitely without the need to discard existing equipment. In using such devices for segmentation of the LAN for the purpose of multimedia applications, it must be kept in mind that latency for such use must be low and constant. A LAN segmentation scheme using an Ethernet switch is illustrated in Fig. 12.4.

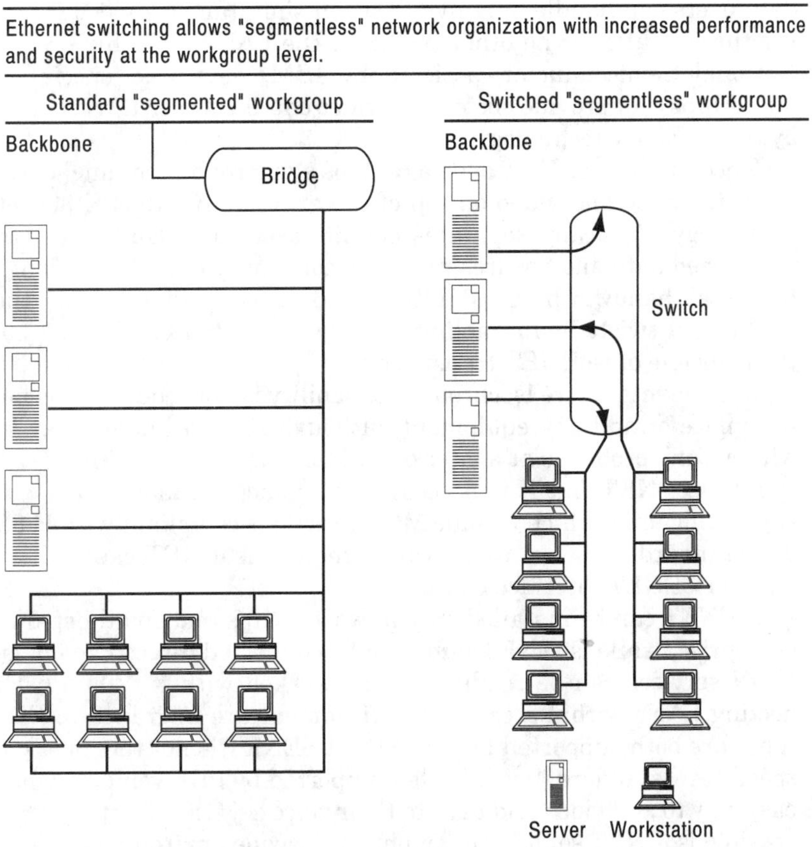
Ethernet switching allows "segmentless" network organization with increased performance and security at the workgroup level.
Standard "segmented" workgroup
Switched "segmentless" workgroup
Backbone
Bridge
Backbone
Switch
Server
Workstation

Figure 12.4 The switched Ethernet LAN. (*Source:* LAN Times, *January 10, 1994, p. 77.*)

Isochronous Ethernet Solution

An alternative solution designed to enhance existing LANs to support two-way, low-latency simultaneous multimedia communications integrating voice, video, and data is the isochronous Ethernet (isoENET). Isochronous capability which delivers signals at a specified, constant rate is desirable for all time-dependent continuous data such as voice and motion video. Unlike transmission with packet-switched technologies, isochronous transmission guarantees timely delivery of information avoiding jitter and delays that are deadly to the intelligibility of voice or video traffic. Isochronous transmissions are ideal for integrated multimedia in real time and can be used like a conventional digital telephone, providing pipelines for two-way, simultaneous real-time, digital communications.

The isochronous nature of video traffic places the heaviest bandwidth demands on existing Ethernet and token ring LANs and internetworks. As a result, a 10-Mbps Ethernet segment that otherwise can support hundreds of data users can handle only five to seven simultaneous video streams, assuming there is little or no other traffic on the LAN. With this problem in mind, National Semiconductor, assisted by IBM, developed special compression microchips that provide six or more video channels over LAN segments enabled by the isoENET technology.

Conceptually isoENET adds a 6-Mbps synchronous channel designed specifically for voice and video on top of the existing 10-Mbps Ethernet. IsoENET technology physically separates circuit-based video and voice streams from packetized data and assures real-time continuity of transmission. As a result, the total bandwidth is 16 Mbps, and it is available over the standard unshielded twisted-pair cabling already in existence. Figure 12.5 illustrates the principle of isoENET transmission.

On segments where isochronous capability is not required it is not necessary to replace or add any equipment, although such segments may be upgraded without any problems at a later date. This minimizes the initial cost of conversion to isoENET, but it is necessary to replace standard hubs with isoENET supporting units which include MCU functions compliant with H.320, 231, and 231 standards. User stations which require isoENET capabilities must also install isoENET interface boards.

IsoENET can be installed only in workgroups and small departments in an enterprise, while isoENET hubs can be connected directly to one another and ISDN services can be readily internetworked with it. When backbones connecting LANs within a campus environment are upgraded to FDDI or ATM, these are both supported by isoENET. IsoENET is not the most popular high-speed LAN standard, but it is also supported by PBX vendors, who see it as an easy way to add video and data to their voice services. Figure 12.6 illustrates a possible isoENET solution in an enterprisewide environment.

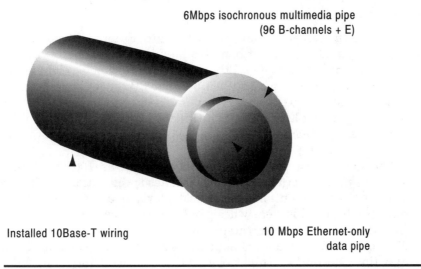

6Mbps isochronous multimedia pipe
(96 B-channels + E)

Installed 10Base-T wiring

10 Mbps Ethernet-only
data pipe

Figure 12.5 The isoENET transmission principle. (*Source:* LAN Times, *March 14, 1994, p. 7.*)

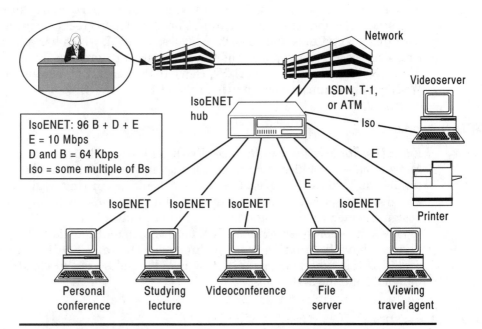

Network

Videoserver

IsoENET
hub

ISDN, T-1,
or ATM

Iso

IsoENET: 96 B + D + E
E = 10 Mbps
D and B = 64 Kbps
Iso = some multiple of Bs

E

E

Printer

IsoENET   IsoENET   IsoENET   IsoENET

Personal
conference

Studying
lecture

Videoconference

File
server

Viewing
travel agent

Figure 12.6 The isoENET network solution. (*Source: National Semiconductor promotional materials.*)

Fast Ethernet

Fast Ethernet is an evolution of standard Ethernet technology which can deliver up to 100 Mbps of bandwidth over a twisted-pair wiring in the star topology configuration. Fast Ethernet is a new technology whose specifications were completed only in mid-1993. Because Ethernet is by far the most widely used LAN standard, there is considerable interest in Fast Ethernet, which is based on existing Ethernet specifications. There are two Fast Ethernet standards that have emerged so far, and this appears to be a drawback of the concept at this time.

The initial Fast Ethernet standard, known as *100Base-X,* preserves as much as possible of the original Ethernet specifications including compatibility with the contention mode. It is deployed in a star-shaped topology using a central hub and is clearly designed to protect as much as possible the original investment. A consortium known as *Fast Ethernet alliance for 100Base-X* of over a dozen vendors supports this standard, including DEC, Grand Junction Networks, Intel, National Semiconductor, Sun Microsystems, SynOptics, and 3Com.

The second Fast Ethernet standard, known as *100Base-VG,* departs from the original Ethernet specifications and operates on voice-grade lines. This concept includes isochronous features with dedicated bandwidth and security capabilities. As such, it has priority protocols which make it particularly suitable for handling video and multimedia traffic. The 100Base-VG technology also works with token ring LANs and as such is favored by IBM, which has a huge token ring user base. The sponsors of this standard are grouped in the 100VG-AnyLAN Forum and include AT&T, Hewlett-Packard, IBM, Novell, Microsoft, Wellfleet Communications, and Ungermann-Bass.

FDDI Networks

Fiber Distributed Digital Interface (FDDI) is a network protocol offering 100 Mbps of bandwidth designed to be used as a corporate data highway linking smaller and slower LANs such as Ethernet and token ring. FDDI LANs are also used for specific throughput-intensive applications, including CAD/CAM design and medical imaging, where very high bandwidth is needed at the desktop level. In order to streamline FDDI for use with multimedia traffic, two new variants have been developed: synchronous FDDI and FDDI II. In spite of costs, FDDI is experiencing rapid growth and its sales are expected to reach $1 billion during the mid-1990s.

Asynchronous FDDI

Conceptually FDDI is a dual, counterrotating token ring topology, with 100-Mbps bandwidth similar to the token ring LAN in which a token is circulated from station to station and transmission can occur only if a station is in possession of a token at the time. As a result of the dual rings, FDDI is highly

With its 100-Mbps bandwidth, FDDI can readily serve as a backbone for multiple LANs and servers. And since this token-passing technology is implemented with dual fiberoptic rings, it's highly fault-tolerant.

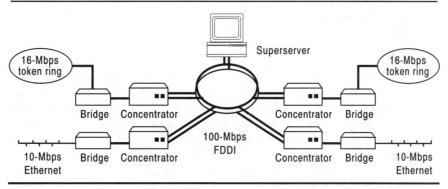

Figure 12.7 FDDI network architecture. (*Source: Reprinted from* Data Communications, *August 1993, p. 81, copyright by McGraw-Hill, Inc., all rights reserved.*)

fault-tolerant because faults occurring on the primary ring are automatically bypassed at specific stations which are always linked to both rings. Figure 12.7 illustrates a typical FDDI network architecture.

FDDI is of value linking multimedia servers to individual stations for store-and-forward transmissions because it has sufficient bandwidth for such traffic. However, FDDI networks process time-sensitive multimedia traffic properly only if asynchronous traffic is restricted on the network. This results from the fact that FDDI networks do not contain prioritization features, and time-sensitive multimedia traffic will exhibit serious delays when network usage goes up.

The synchronous FDDI alternative

While asynchronous FDDI networks can handle time-sensitive multimedia traffic, if their asynchronous traffic volume is severely restricted, synchronous FDDI is designed specifically to provide priority to multimedia traffic on the network. A synchronous FDDI network is configured with a certain proportion of bandwidth dedicated to time-sensitive multimedia traffic by separating stations with multimedia requirements from the rest. This has the effect of prioritizing synchronous traffic such as audio and video for these stations. Since the protocol is based on the original FDDI standard, many asynchronous FDDI adapters can be upgraded using software enhancements. This is an important consideration because it permits introduction of multimedia capabilities into the networks and protection of existing investments in FDDI LANs. Such upgrades, however, depend on availability of adequate memory in the hardware before this option can be implemented.

A synchronous FDDI Forum consisting of 16 vendors of FDDI products is looking into the potential for such products, but many vendors are reluctant to become involved with such products because there is already a higher-quality FDDI II network being developed which has superior characteristics with regard to multimedia communications.

The FDDI II option

FDDI II is a whole new standard with an architecture which differs from the token (ring) passing topology of the other FDDI options, and it offers superior network latency for time-sensitive multimedia traffic. It is based on a multiplexing approach where the 100-Mbps bandwidth is split up into 16 channels, each of which can be further subdivided and allocated to asynchronous or isochronous traffic based on user demand. FDDI II can provide precise delivery times for multimedia transmissions, eliminating the fluctuating and unpredictable delays that occur with other FDDI solutions. Figure 12.8 illustrates the FDDI II multiplexing approach and access control principle that distinguish it from the original FDDI formats.

However, FDDI II is not compatible with the other FDDI standards and requires costly hardware additions, including the media-access-control (MAC) adapters and multiplexing equipment. On the other hand, its node-to-node latency is in the order of 125 μs, which is almost two orders of magnitude less than the delays on the standard FDDI networks and well below the 100-ms level considered tolerable by real-time multimedia end users. The FDDI II option offers a logical migration path to ATM in the future, and many believe that all FDDI solutions are temporary until such time that more cost-effective ATM products come to market.

The PBX Solution

Relatively little known and practiced so far is the *private branch exchange* (PBX) solution to multimedia networking. A number of PBX equipment manufacturers have upgraded their products to handle multimedia traffic. This applies primarily to circuit-switching and hybrid models of PBXs that can carry video transmissions using LAN connections or dedicate one channel to voice and video and another to bursty datastreams. Another approach is to transmit voice and video traffic through the PBX and data over Ethernet LAN, but this means integration of the two streams at the receiving end. Nevertheless, since PBXs are installed in many organizations if there is an opportunity to network multimedia traffic through them at a marginal cost it should be explored.

PBXs are well suited to carrying time-sensitive video traffic because they provide a guaranteed maximum latency, but their circuit switching technology offers limited bandwidth capabilities, which results in poor-quality video images. All the multimedia transmission schemes available from PBX ven-

Network adapters conforming to the forthcoming FDDI II spec will feature separate media access control (MAC) circuitry to handle isochronous traffic and conventional asynchronous traffic, shrinking isochronous circuit delay to 125 microseconds.

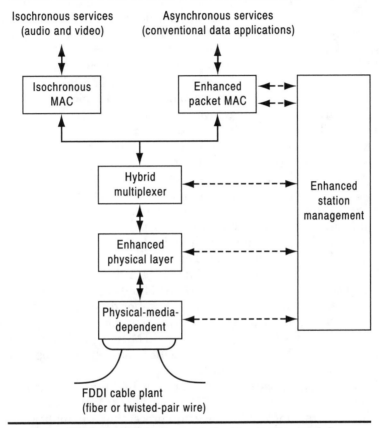

Figure 12.8 FDDI II multiplexing network. (*Source: Reprinted from* Data Communications, *July 1993, p. 56, copyright by McGraw-Hill, Inc., all rights reserved.*)

dors are proprietary because there are no standards defining multimedia traffic on customer premise. Although some vendors support 100-Mbps fiber connections in campus networks made up of distributed PBXs, these may not be sufficiently compatible with each other to support interactive collaborative sessions.

Interestingly, PBX vendors themselves are adding ATM functionality to their products in form of adjunct ATM switches. These can handle high-bandwidth traffic including data, multimedia, and high-definition video, while the original PBX will handle voice traffic. In some instances native ATM PBX products are being designed to carry all traffic.

ATM Networks

The asynchronous transfer mode (ATM) is seen as an ideal method for handling multimedia traffic because it can transmit any mix of packetized or time-sensitive data at very high speeds. ATM can connect with data sources and sinks at any level in the network and can be extended if necessary to end users themselves. Because it transmits data without degradation for thousands of miles and is relatively expensive as a new technology, ATM is seen at present as the best solution for internetworks and WAN applications.

ATM is based on a technique known as *cell switching,* which converts all digital data types into fixed-length cells of 53 bytes each. Since these cells are very small and identical in length, they can be switched rapidly by hardware alone. This gives ATM an advantage over methods which are considerably slower because they depend on hardware and software switching technologies. In an ATM network LANs and WANs carrying any type of voice, data, or video traffic can interoperate seamlessly and in real time.

The concept of ATM switching is rather elegant and quite simple. An ATM interface accepts streams of data from different sources such as workstations, LANs, WANs, and other devices at various bit rates and divides them into cells of equal length. ATM cells consist of 48 bytes of data and 5 bytes of control information and are routed through high-bandwidth trunk lines at speeds of 150 or 600 Mbps and even higher. The ATM switch can be compared to a funnel that accepts and slices up variable-length data packets, video, audio, or other ATM cells into a multiplexed stream of cells at a much higher speed. The cells are reassembled into the original packet or datastreams at the other end of the process. Figure 12.9 represents this concept in form of a diagram. As a result, a multimedia transmission in an ATM network becomes one or more streams of data, each dedicated to a specific bandwidth to support voice, video, data, or graphics as the case may be.

ATM is making inroads into workstation LANs and backbones, in public MAN and WAN services, and in WAN backbones. Visualization, multimedia, video, medical imaging, and transaction processing are applications that benefit the most from ATM. The ATM Forum founded in October 1991, which includes over 250 member firms, promotes interoperability between ATM vendors and is working on specifications for standards such as user-network interfaces (UNI) and other parameters.

As ATM usage increases and prices drop, such devices will become cost-effective for individual users. This is an important consideration because ATM offers the infrastructure that can bring about the virtual LAN, which is considered by many corporate planners the ultimate in networking technology. ATM is an expensive method at present, but it is expected to become the best solution, eventually making virtual LANs and interactive enterprise a practical reality.

An ATM switch takes all incoming packets, divides them into 53-byte cells, adds headers, and then retransmits them onto the network. If SONET is being used, the available bandwidth has already been divided into slots.

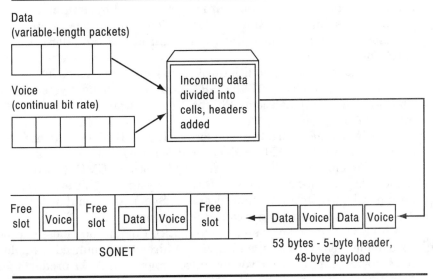

Data
(variable-length packets)

Voice
(continual bit rate)

Incoming data divided into cells, headers added

| Free slot | Voice | Free slot | Data | Voice | Free slot |

| Data | Voice | Data | Voice |

SONET

53 bytes - 5-byte header, 48-byte payload

Figure 12.9   The ATM technology. (*Source: Reprinted from* Data Communications, *March 1993, p. 91, copyright by McGraw-Hill, Inc., all rights reserved.*)

The Virtual LAN Concept

Virtual networking is defined as the ability to accommodate mobile individual users among collaborating groups on a LAN without the need to make any changes either to the desktop, workstation, portable computer, or any cabling or software. This means that instead of rewiring the LAN infrastructure or creating temporary LANs to support a new project team, virtual networks permit to pull together a logical workgroup regardless of the location or the type of LAN connections. It is the dream LAN for the proponents of interactive enterprise and the virtual corporation models.

Virtual networks raise a number of management issues pertaining to network security between mobile or newly created groups. They are also hard to represent as a physical infrastructure because of the changing users and configurations. As a result, there is concern that LANs that may evolve to span the globe will become unmanageable.

Virtual LANs can be implemented over CDDI or FDDI backbones with Ethernet switching in the short term, but in the long run ATM technology is expected to be the mainstay of these networks.

Vendors of Multimedia LAN Products

Most LAN vendors can reconfigure a network to provide the microsegmentation required to handle multimedia traffic. The key to this conversion, however, is the segment switching hub that is now supplied by a number of small companies. These include Alantic, Artel Communications, Bytex, Cabletron Systems, Kalpana, Lannet Data Communications, Standard Microsystems, and 3Com.

PBX vendors with capabilities or kits to convert switches for handling multimedia traffic include AT&T, GPT Communications Systems Ltd., NEC America, Northern Telecom, Philips Communications Systems, Rolm, and Siemens AG.

Aside from major hardware vendors including IBM, over a dozen small ventures specialize in FDDI products and services. These include Alpha, Inc., Ascom Timeplex, Cabletron Systems, Chipcom, CMC Network Products, Crescendo Communications, David Systems, Interphase, Network Peripherals, SynOptics Communications, Synernetics, and 3Com.

As far as ATM switching is concerned, major suppliers of LAN products such as Chipcom, Cabletron Systems, SynOptics Communications, and 3Com are developing ATM products, and IBM has already announced a whole array of ATM devices. Many other vendors also announced ATM products, and more than 100 are believed to have them in development. A number of new ventures are making a name for themselves as ATM specialized suppliers, including Adaptive, Fore Systems, and Newbridge Networks.

13

Multimedia Internetworking

Internetworking means collaboration between users within one LAN with those on other LANs within an enterprise campus, a metropolitan area, or in different WANs, that are themselves interconnected through private and public lines throughout the world.

It can be readily demonstrated that a LAN can handle multimedia transmissions in an isolated environment. But the more important question is whether multimedia applications, particularly time-sensitive group conferencing and video distributions, can coexist on the same network with traditional traffic without severely affecting the performance of the systems. This issue is further aggravated when LANs are internetworked with other LANs within and without the enterprise.

The broadband backbones interconnecting those LANs may have considerable more bandwidth than the LAN segments, but these are designed primarily to handle the growing data traffic generated by rapidly increasing numbers of users, more complex applications, and increasing output of more powerful workstations.

Multimedia LAN Interconnecting Issues

Introduction of interactive multimedia communications within a single LAN is not a trivial task, but the major issue is one of sufficient bandwidth, which can be resolved in a number of ways as outlined in Chap. 12. Within a single LAN multimedia transmissions are made over relatively short distances and across a limited number of interfaces in a hub or switch depending on the topology.

Internetworking introduces additional barriers to the flow of multimedia traffic in form of bridges, routers, repeaters, and other hubs, each contributing to the cumulative delays between two or more users collaborating from several different LANs and communicating across backbones, WANs, and other services.

The question of acceptable latency between such users becomes the paramount issue which governs response time of all LAN users and determines the throughput of the networks and the quality of communications, particularly where real-time audio and video transmissions are involved.

Time Sensitivity of Multimedia Traffic

As discussed previously, *audio and video traffic* represents time-sensitive multimedia traffic without which telecollaboration in form of multipoint conferencing and other interactive group exchanges cannot take place. Some collaborative multimedia activity will take place within a single LAN, particularly if it is a large network. But it is collaboration between individuals on various LANs and outside the enterprise that are strategically more important. This is a natural development toward the interactive enterprise as such and positioning it to participate in future virtual corporate operations at very short notice.

In conventional client-server networking, throughput can be measured by the number of transactions per unit time and response time by the average response time per transaction. With multimedia networks, two additional independent yardsticks must be added to network management to measure the effect of delay on perception of audio by the human ear and the number of video frames transmitted per second and its effect on the human eye.

Interactive multimedia communications means real-time transmission of audio and video to multiple participants on various LANs. It also means real-time responses between any and all parties involved. This aspect of interactive multimedia collaborative sessions calls for faster-than-real-time (FTRT) processing of multimedia transmissions, which in themselves contain time-sensitive audio and video traffic. Such requirements for time-sensitive processing puts unusual demands on most existing LANs and WANs. Nevertheless, successful development of internetworking connectivity with acceptable levels of multimedia communications is a critical milestone on the road toward interactive enterprises and their participation in virtual corporations of tomorrow.

A Question of Acceptable Latency

Latency, which has been dubbed "the silent killer" of internetworks, can be defined as the delay that is incurred as digital data or a unit of data such as a packet or frame passes through a device in a network. Even though latency has a tremendous impact on network performance, vendors seldom refer to it as a measurement of the capabilities of their products. Yet latency is the main cause of what is sometimes known as "networking in slow motion."

Latency is much more pronounced in enterprise networks where every bridge, router, and hub contributes to the cumulative effect. It is also a very frustrating performance issue to handle. When a device is not powerful enough to process a given traffic level, the load can be reduced by providing more devices, more powerful units, or both. This does not necessarily reduce any latency problems which may exist even with relatively lightly loaded networks.

Recent tests performed on bridges and routers by independent testing laboratories suggest that latency is responsible for reducing native equipment throughput to anywhere between 56 and 78 percent depending on a particular product. The more such devices exist on the path between two LAN users who are internetworking, the more latency comes into play. This means that interactive multimedia collaboration between users on LANs that are physically separated by many steps may be difficult to achieve if the cumulative latency is too high.

Some applications have a high tolerance for latency because their transmission is not time-sensitive even though these may be high-resolution images or visualizations. Transmission of such traffic may be scheduled during low usage periods on the networks. Other applications, notably audio and video regardless of their quality, have a very low tolerance of latency with audio being the least forgiving. Table 13.1 illustrates latency tolerance of some typical applications.

The audio latency factor

The quality of digitized audio is a very subjective phenomenon, and acceptance also depends on a specific application. However, in collaborative multimedia it is mandatory that any voice transmissions be clear and intelligible to all parties at all times. Telephone voice is not very bandwidth intense, but digitized stereo music requires about 0.5 Mbps of bandwidth. What is desirable in collaborative multimedia is audio digitization and compression that produces a good-quality voice transmission on the poorest communications link in a network within the range between the phone speech quality and stereo music.

The critical issue in interactive audio transmission within a network, however, is not bandwidth but latency and jitter. Once a person starts to speak, intermittent delays during such a session are very noticeable and often annoy-

TABLE 13.1   Latency Tolerance of Typical Applications

Application	Type of transmission	Latency tolerance	Internetworking protocol
High-quality voice and video	Isochronous	Very low	ATM
Low-quality voice and video	Isochronous	Very low	IsoENET, FDDI-II
Transaction processing	Small files	Low	ATM
Medical imaging	Large files	High	FDDI, ATM
Visualization	Very large files	High	FDDI, ATM
Publishing	Large files	High	Fast Ethernet, FDDI
File transfer	Variable	High	Fast Ethernet, FDDI

TABLE 13.2   Audio Latency and the Human Ear

Audio latency, ms	Effect of delay on human voice perception
>600	Speech is unintelligible and incoherent
600	Speech is barely coherent
250	Speech is annoying but comprehensible
100	Imperceptible difference between audio and real speech
50	Humans cannot distinguish between audio and real speech

ing. Thus it is desirable that once a voice transmission has commenced, it should continue uninterrupted.

Table 13.2 shows that the maximum tolerable latency for speech is 600 ms and above that level speech becomes incoherent and incomprehensible. However, experience with satellite communications indicates that a delay of even 250 ms is annoying even though the speech is completely coherent. For quality music, which may have a place in promotional multimedia networks, the latency is more noticeable and any delays must be even less than that.

As far as interactive speech is concerned, network analysts believe that an end-to-end latency should not exceed 100 ms. It is important to remember that multimedia transmissions that do not include interactive speech as a component can tolerate somewhat larger latencies. However, if efficient collaborative multimedia systems are the objective, interactive speech latency limits should be used as a critical measure of performance objectives.

The video frame rate factor

Digital video represents a different set of latency issues, although the human eye is more forgiving than the human ear. Nevertheless, latency is very critical in interactive video when multiple sessions are under way simultaneously and must be synchronized. Codecs involved in video transmission contribute to the delay during transmission and reception of signals. Codecs that provide MPEG compression may contribute up to 30 ms of delay, but some p*64 codecs may have much larger one-way latencies on the order of 100 ms. As with audio, end-to-end latency of 100 ms is acceptable and tolerable.

It is important to remember that the end-to-end transfer delay is a cumulative factor which includes the queuing delay in transmission, propagation delay, and network access delay, which in a multiuser LAN may be the most critical component of the overall latency factor.

Video transmission itself is more flexible than audio because it is two-dimensional and it is possible to trade off the frame rate with size of image within any given bandwidth. An electronic meeting between participants on LAN-connected desktops, ISDN-linked home telecommuters, and phone-line-linked portable units from a hotel room can take place, but the size and quality of the video will vary significantly in each case.

TABLE 13.3   Video Frame Rate Effect on Human Eye

Frame rate, fps	Effect of frame rate on video perception
<10	Motion is not perceived by the human eye
12–15	Jerky motion is perceived by the human eye
30	Broadcast TV-quality motion
60–75	HDTV-quality motion, high-speed motion not blurred
90	The limit of human eye perception
1000	Scientific video quality

Video should always try and maintain at least 15-fps frame rate because this is the limit at which the eye perceives jerky motion. (Table 13.3) The prime objective is therefore to maintain video at over 15 fps by adjusting the size of the image until it becomes very small. This is acceptable when simply observing other participating collaborators but will be inadequate to convey details of an inspection tour of a new facility or moving parts of a new mechanism. Some multimedia compression devices automatically reduce the image size and decrease the frame rate as available bandwidth is reduced until the image is too small and too jittery to be of much use.

Response-Time Requirements

As users and transactions increase within a network, response time increases exponentially until it becomes unacceptable. However, in conventional text and data networking response time may also be a function of cost of operations. It may increase from 2 to 5 s before becoming unacceptable, but it still remains tolerable. This, of course, is not the case with time-sensitive multimedia communications.

Underestimation of potential traffic in client-server environments is quite common, and in the case of designing multimedia networks which will handle unpredictable loads associated with an interactive enterprise or a virtual corporation, it is difficult to avoid. These applications are much more complex, random, and difficult to anticipate than order entry and client billing.

Traditional network designers concentrate on individual component speeds, capacities, connectivity, and migration path potential. In the case of multimedia networking, bandwidth and its capacity to provide real-time response to time-sensitive transmissions is the more critical consideration. Figure 13.1 illustrates the three parameters (response time, throughput rate, and saturation range) and their relationship to throughput of the network. What is needed is a control system that will alert network designers that the bandwidth and latencies of the system are tending toward a level that will create unacceptable audio and video traffic.

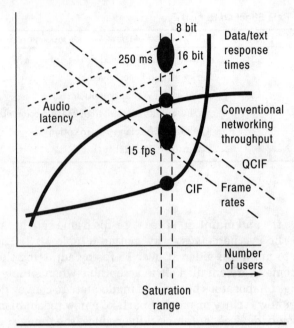

Figure 13.1 Multimedia network transmission parameters.

Throughput Requirements

In the client-server environments in which many LANs operate the distributed nature of interactivity significantly affects response time for all network users. A distributed physical layout which includes a collection of hardware, software, and connectivity devices can increase a network overhead by 10 percent quite easily even in conventional text and data transmitting LANs. Interactive multimedia capabilities increase the complexity of any LAN immeasurably. This is the result of adding special videoservers, compression devices, videocameras, speakers, microphones, digitizers, and associated software. All these generate additional operating overhead during transmission which pertains to housekeeping, security checks, saving of random multimedia data, storage, and backup activities.

The situation becomes very complex where legacy systems with data-center-based networks are collaborating with client-server LANs through a conglomerate of gateways, protocol converters, bridges, and routers, all of which have been accumulated over time rather than having been planned to provide a reasonable response for a constantly growing user population.

Line Quality Factors

Internetworking within an enterprise will most likely depend on a fiberoptic link such as FDDI or Fast Ethernet, but beyond the campus there are WAN services that are available with different data rates and varying levels of service quality. These services range from POTS, through T-1, T-3, ISDN, to SONET but such services are not universally available. As a result, it is improbable that collaborative multimedia sessions can be set up and gotten implemented when needed to all remote areas and across all international borders.

Most analog lines throughout the world are gradually being replaced by digital lines. Eventually digital lines coupled with high-speed digital switches like ATM products will offer public network services that will be suitable for different transmissions including multimedia audio and video traffic with constant quality of transmission regardless of distance. Until such time collaborative multimedia operations will be introduced selectively where cost and quality of service permits.

Multimedia Wide Area Networks

There is increasing concern about WANs for multimedia traffic even though interactive multimedia enterprises are still few and far between. There is a great awareness among forward-looking business executives about profound changes taking place in the socioeconomic patterns of contemporary business environments which are conducive to collaborative multimedia communications. This is best seen in the continuous decentralization of large businesses. A recent survey of Fortune 1000 firms in the United States revealed that on the average such companies operate 234 sites with fewer than 20 employees in 54 percent of them. Collaborative multimedia communications are an ideal medium to increase productivity and competitiveness of these organizations. It is therefore inevitable that cost-effective high-speed WAN services must come into being to make wide area collaborative multimedia communications a reality.

A recent study by the Yankee Group of Boston investigating the internetworking plans of large Fortune 1000 companies revealed that most plan to transmit imaging, videoconferencing, and multimedia traffic over WANs within the next few years. In 1993 most WAN traffic of the large corporations consisted of conventional transaction processing, E-mail, and LAN-to-LAN communications. At that time only 18 percent of representative users used WANs for videoconferencing, 16 percent for imaging, and 12 percent for multimedia traffic. However, use of WANs for such transmissions is expected to increase dramatically by 1998. It will triple for videoconferencing to 56 percent of corporate users, increase almost fivefold to 72 percent of users for imaging applications, and grow even faster to include 64 percent of corporations for all other multimedia traffic. A typical WAN architecture is illustrated in Fig. 13.2.

Cost and Availability of High-Speed WANs

The greatest challenge in multimedia is to develop a cost-effective WAN service because high-speed low-cost WAN services are critical to multimedia commu-

Newbridge Networks Vivid hardware includes ATM hubs, access switches (Ridges), and LAN Service Units. Route Server software residing throughout the internetwork works with Ridges to provide dynamic routing of Ethernet and token ring traffic.

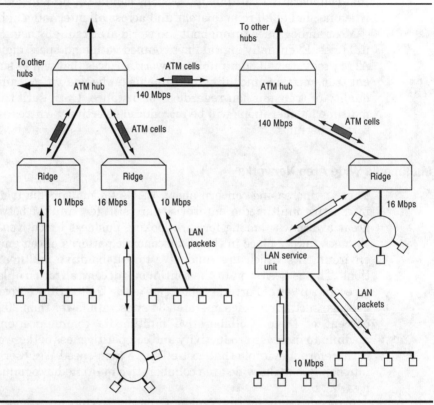

Figure 13.2 Typical WAN architecture. (*Source: Reprinted from* Data Communications, *March 1993, p. 41, copyright by McGraw-Hill, Inc., all rights reserved.*)

nications. Despite some efforts on the part of some carriers to further develop ISDN and introduce SMDS on a national basis, WAN internetworking at present remains a bottleneck to faster development.

WAN transmissions can be provided by a number of existing services, few of which present cost-effective alternatives for multimedia traffic. These include leased circuits, telephone lines, ISDN services, or public and private data networks. Bandwidth of these WAN communications services vary from 19.2 Kbps for the typical analog telephone all the way to 2.488 Gbps for SONET-caliber service. The other problem with those services is the fact that they are not uniformly available in all parts of the country, let alone the world.

Nevertheless, WANs and MANs are critical to multimedia networking if the advantages of collaborative conferencing and interactive enterprises are to be realized. As a result, interoperability is now an important issue and there are moves to develop various alternative transmission facilities and schemes to facilitate multimedia communications over long distances and make them more cost-effective. These include new forms of services, intranetworking hubs, and integrated access devices (IADs) that use reverse multiplexing to combine the bandwidth of up to four T-1 circuits to provide a 6-Mbps bandwidth transmission channel.

The best course of action prior to the availability of high-speed cost-effective WAN facilities is to prepare for the eventuality by developing LAN architectures that are conducive to collaborative multimedia communications. This means a support of two-tiered star topologies in existing or newly developed LANs because it is the most useful for the implementation of microsegmentation on a user-per-user basis. The Ethernet 10Base-T LAN standard offers the lowest cost alternative because it is the most popular, and Fast Ethernet links with 100-Mbps bandwidth will use the same cabling and interfaces that are already in place.

Nevertheless ATM product prices are also falling, and the Yankee Group forecasts a more rapid drop in ATM prices per 155-Mbps connection than those of competing 100-Mbps FDDI products. Another study of ATM equipment prices by Ryan Hankin Kent projects that by 1995 an ATM network node including adapter cards and hub port will drop to about $85 per Mbps compared with $190 per Mbps for token ring and $240 per Mbps for Ethernet despite the fact that ATM was the most expensive technology at the end of 1993.

High-Speed Interconnections

The only high-speed WAN collaborative multimedia solutions that can handle the real-time audio and video traffic involved are FDDI and Fast Ethernet networks with 100-Mbps bandwidth and the ATM switches and networks that can expand from 45 Mbps all the way to 155 Mbps and eventually the 2-Gbps range. It is questionable whether such high bandwidth will be necessary for handling collaborative multimedia, but ATM will offer bandwidth on demand, and this is its real attraction as well as the capability to handle all forms of data simultaneously.

It is also important to keep in mind the overall architecture in WAN interconnectivity to make sure that cumulative latencies do not adversely affect the audio and video transmissions across the WAN. It is deceptive in highly segmented networks with numerous concentrators, FDDI backbones, bridges, and routers to depend on demonstrations of collaborative multimedia communications feasibility unless these are made under a heavy network loading condition.

It is much more advisable, if at all possible, to plan multimedia WAN connectivity from the outset. In cases of multiple segments the best solution is to create a hierarchical network architecture which consists of switches and intelligent hubs controlling individual LANs through a two-tiered tree topology. Under such conditions connectivity between users in distant LANs is established through an intelligent hub and a switch bypassing all other LAN segments and reducing the overall latency to an acceptable level. Figure 13.3 illustrates this type of WAN arrangement.

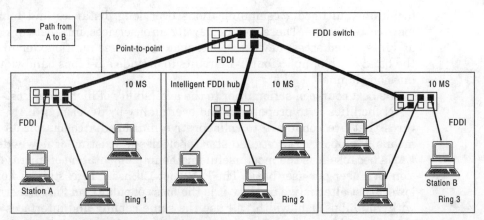

Figure 13.3 Hierarchical network architecture.

Asynchronous Transfer Mode Promise

The most promising technology for high-speed internetworking with collaborative multimedia communications is undoubtedly ATM, the underlying concepts of which are discussed in some detail in Chap. 12. ATM has an advantage over other competing products because its cell switching is hardware-based and as such is three orders of magnitude faster than software-based processing of standard routers. This also creates a problem when the load on an ATM network reaches 70 percent of capacity, but congestion-control schemes are already being developed under the guidance of the ATM Forum.

Increasing sophistication of networking applications and inclusion of real-time audio and video make ATM practically the only technology that can support all the data types. ATM is expanding in three specific market segments. These include corporate LANs and backbones, infrastructures for value-added networks (VANs) services as these expand to include multimedia data types, and public networks based on ATM technology.

ATM backbones offer several advantages over FDDI and other solutions. They offer greater bandwidth and allow the creation of virtual workgroups, thus eliminating the need to reconfigure switches when personnel moves to different locations. ATM backbones assist in easing bottlenecks created on client-server connections because they offer 155-Mbps switched connectivity. Figure 13.4 illustrates an ATM backbone with Ethernet switches and LANs.

A nationwide ATM service accessible at rates as low as 1.544 Mbps has been announced by AT&T Interspan at rates comparable to T-1 lines. Although initially this service handles only data and constant-bit-rate video, integration of voice, video, and data is expected in the near future.

Major Products and Service Suppliers

Traditional WAN products and services include interfaces, routers, bridges, gateways, hubs, and various services that provide the equipment to intercon-

By attaching existing LAN segments to switching hubs, managers can provide video capabilities to more users. For campus networks, FDDI can be used instead of asynchronous transfer mode (ATM) as the backbone network.

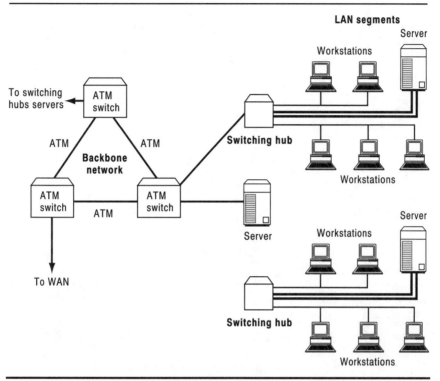

Figure 13.4 ATM backbone WAN solution. (*Source: Reprinted from* Data Communications, *January 21, 1993, p. 62, copyright by McGraw-Hill, Inc., all rights reserved.*)

nect with WANs and monitor the performance of LANs in relation to each other. Most of those products, however, pertain to conventional text and data networking.

In the area of special equipment such as hubs and IAD devices, Larse Corporation, NEC America, and Xyplex already offer new products to handle multimedia transmissions.

One of the most interesting areas is the whole array of ATM capable products that are being developed by a number of network equipment manufacturers. These include companies such as Agile Networks, AT&T, Alantec, ADC Kentrox, Cisco Systems, Digital Link, Fore Systems, Lannet Data Communications, Motorola Codex, Stratacom, Synoptics, and 3Com.

14

Multimedia Network Servers and Databases

Multimedia client-servers, databases, and massive storage systems are critical elements of multimedia networking infrastructures without which collaborative multimedia communications are impossible. Although much attention is given to such issues as time sensitivity of multimedia traffic, high-speed bandwidth requirements, and LAN-WAN interconnectivity, relatively little is mentioned about storage and retrieval of multimedia objects and traffic and the associated hardware and software technologies.

In standalone multimedia platforms the issue is relatively simple. Multimedia objects or applications are stored on CD-ROM drives and retrieved when needed through decompression devices for delivery on PC screens. Many conventional videoconferencing systems are also regarded as videotelephone sessions with little or no requirements to save the proceedings aside from recording some on videotape. In these cases the interactive multimedia activities do not have a major impact on storage and manipulation of multimedia objects or data types.

The situation undergoes a radical change when interactive multimedia networking comes into play. This is particularly true of collaborative multimedia communications, which require access to various sources of multimedia objects as well as capabilities to display and manipulate such objects in real time during multimedia conferencing sessions. In addition, because of the limitations of acquiring all the desirable participants at any one time, there is a distinct need to save and retrieve at a later date some of those sessions and retransmission with changes and comments of other parties. Storage requirements of some representative multimedia objects are summarized on Table 14.1.

Similarly, in the consumer markets there is a demand for specialized videoservers and sophisticated billing systems that can retrieve, transmit, dis-

TABLE 14.1   Multimedia Objects Storage Requirements

Object type	Size and bandwidth
Text—single page	2 kbytes
Simple image—uncompressed	64 kbytes
Detailed image—uncompressed	7.5 Mbytes
Voice or telephone—8-kHz, 8-bit sampling mono	6–44 Kbps
CD audio—44.1-kHz, 16-bit sampling stereo	176 Kbps
Animation—320 × 240 × 16-bit color frames at 16 fps or image and synchronized audio at 15–18 fps	2.5 Mbps
Video—TV analog 640 × 400 × 24-bit color at 30 fps or digital image and synchronized audio at 24–30 fps	27.7 Mbps

play, and account for randomly chosen multimedia services with which large numbers of customers may want to interact electronically. All these multimedia networking concepts impose massive storage and data manipulation demands on any existing or future networking systems and transmission facilities.

Massive Storage Requirements

Multimedia data processing imposes truly awesome data storage requirements on a network. Multimedia storage must provide massive capacities measured in terabytes (Tbytes) with capabilities to manipulate gigabyte (Gbyte)-range data objects at a time. This results from the fact that a single digitized video frame consumes 1 Mbyte and 1 h of compressed video requires 1 Gbyte of storage. Such massive storage must provide random access to many simultaneous users and an efficient set of tools to revise and update all the content. A critical characteristic of multimedia storage is speed of data retrieval, keeping in mind applications where audio and video data cannot tolerate delays above certain practical limits. If such a multimedia storage system is required to manage voluminous video files, this entails the use of very fast processors with a lot of data transfer bandwidth capability to handle video traffic in an efficient manner.

Such demanding storage requirements suggest that multimedia databases reside on very large magnetic disks or RAID arrays, which are now increasing in popularity. The IBM midrange AS/400 system is often used as a LAN storage server because it can handle up to 123 Gbytes of DADSD storage and up to 280 Gbytes of optical storage as well. Mainframes are also offered as multimedia servers for enterprisewide applications. Figure 14.1 illustrates a typical LAN storage server arrangement.

Vast quantities of multimedia materials also exist in analog form, and there is a definite need to develop availability of such items through extended stor-

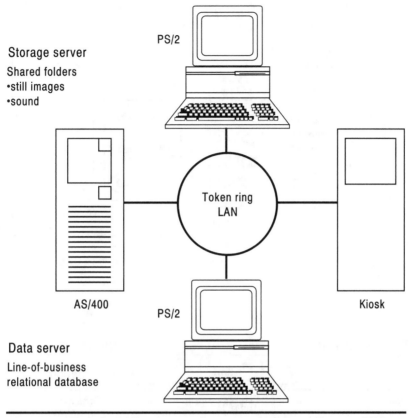

Storage server

Shared folders
•still images
•sound

PS/2

Token ring
LAN

AS/400

PS/2

Kiosk

Data server

Line-of-business
relational database

Figure 14.1 Typical LAN storage server arrangement. (*Source: IBM Corporation.*)

age hierarchies. These must support various devices such as VHS tape drives, CD-ROMs, DAT tapes, and optical disks. Such storage devices cannot be directly on-line to support interactive multimedia operations, but automated means must exist to access, inspect, and digitize such content should it be required for development of multimedia applications. This concept is illustrated in Fig. 14.2.

An associated issue with multimedia storage systems is the question of connections and cabling required to transfer efficiently massive volumes of multimedia data. Software tools to manage, evaluate, and manipulate multimedia materials must also be acquired, many of which take the form of multimedia file managers and multimedia databases.

Multimedia Databases

Multimedia databases are basically software programs that can manipulate multimedia data types consisting of text, images, audio, animation, and video

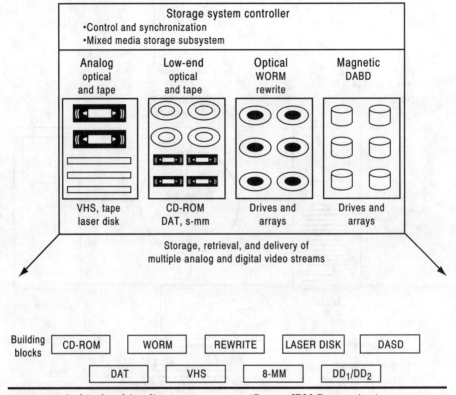

Figure 14.2 Archival multimedia storage concepts. (*Source: IBM Corporation.*)

elements using a workstation or a PC. This objective can be accomplished in one of several ways. Traditional relational database management systems (RDBMs) are being expanded to handle multimedia data types; special object databases (ODBs) are designed to handle complex objects such as BLOBs which include all multimedia data types; or hybrid database solutions which combine RDBMS and ODB concepts. Major RDBMS vendors are developing multimedia capabilities in their traditional products often in collaboration with ODB vendors, but their approach to handle multimedia within RDBMS systems varies from product to product. It is believed that most RDBMS products will eventually include multimedia storage and client-server development capabilities or options.

There is a category of multimedia file management software products that are sometimes classified as multimedia databases. In effect, these are multimedia object file organizers or catalog programs and should not be confused with on-line database systems with multimedia manipulating capabilities.

Multimedia applications impose several functional requirements on database management systems as a result of the necessity to handle very large and complex data types. These present special data representation, data manipulation, data management, and data storage issues. Multimedia databases must be able to handle all these requirements on a timely and efficient basis when operating within a LAN environment.

Multimedia file managers

Multimedia file managers are relatively simple programs that allow users to organize, index, and retrieve multimedia objects electronically or otherwise stored in conventional files. These managers are used to categorize and select multimedia contents using descriptors and indexing for retrieval. Some programs include a panel which displays thumbnail images representative of each object stored while other systems require language support for manipulation. Basically these programs are standalone tools designed for evaluation and selection of multimedia objects that are without relationship to each other or any other data.

The primary objective is to store information about multimedia files including name, size, and type, along with thumbnail image and pointers to their physical location. Many of these file managers allow seamless integration with other applications. All these programs provide multiple ways to view the data either as an array of thumbnail images, a textual list, or as individual previews. Many of these programs include links to popular editing software, while others provide drag-and-drop or cut-and-paste capabilities. The most valuable features of those file managers are fast search and sort, and some include special capabilities to look for colors, texture, and shapes in stored images. Some of the more sophisticated programs include JPEG compression, screen capture, and image editing tools.

Depending on configuration of these products, they are appropriate for standalone or workgroup deployment and normally reside in a file server for multiuser access. Some of these programs claim to support as many as 20 users simultaneously, but generally network performance is drastically degraded when more than five users are involved on a single LAN.

Three types of these multimedia file managers exist. The first and simplest offers a scrapbook utility and seldom include network support. The more sophisticated programs provide off-line file storage on removable cartridges, optical disks, or CD-ROMs. Some manual intervention may be required in working with such files. For workgroups multimedia managers provide networking capabilities which include remote volume mounting and password protection for archival and simultaneous catalog access.

Multimedia database characteristics

Multimedia databases can be implemented using RDBMS, ODB, hierarchical, and flat file database models or even combinations of all those categories. Initial decisions must be made on the basis of existing system platforms, memory requirements, operating systems, file compatibility, query facilities, multiuser features, programmability, and price. Such standard parameters are used to evaluate any type of database, but additional parameters must be taken into account when developing a multimedia database system.

Multimedia applications place very specific demands on a database structure over and above the standard data model and functional requirements. These pertain to extensibility, flexibility, and efficiency of the system when handling multimedia data types. A conceptual multimedia database architecture is illustrated in Fig. 14.3.

Extensibility of the database is required to provide special support for new multimedia related devices. These include cameras, digitizers, special displays and presentation systems, various massive storage systems such as CD-ROMs, WORMs (write-once read-many disk drives), optical jukeboxes, videotape libraries, and a specific framework for logical access to such devices. Users should be able to select specific compression ratios and trade them off against quality, size of video, or frame rate. These are not standard features of a typical database and must be developed specifically for multimedia databases. For

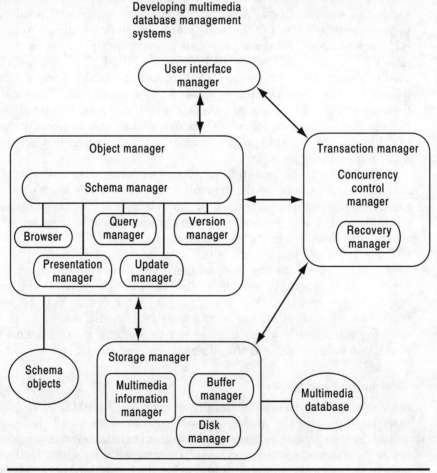

Figure 14.3 Multimedia database architecture. (*Source: McGraw-Hill DATAPRO, Multimedia Solutions.*)

efficient storage and manipulation, multimedia data types and BLOBs also require compression and special formatting features.

Flexibility in a multimedia database management system means provision for storage and manipulation of spatial and linear multimedia objects that may or may not be persistent in nature. Spatial multimedia objects such as images should be addressable in segments chosen by the user. Linear multimedia objects such as audio or text may have to be specified in units of time for presentation. If an image is changed during manipulation and then returned to storage, it is a persistent multimedia object which must be updated as such by the system in all its instances. In addition, a multimedia database must be able to provide controls for capture and display of multimedia data comparable to videotape devices which include functions such as pause, continue, rewind, fast forward and backward, and stop which are critical for handling time-sensitive multimedia transmissions.

Efficiency of operations of a multimedia database is extremely important because multimedia data types or BLOBs are very large objects. It is therefore necessary to provide a means of reducing storage requirements whenever possible either through compression or by sharing common storage space. Elimination of unnecessary duplication and buffering of multimedia data within the system can also optimize the transfer of BLOBs between capture devices, storage systems, and presentation facilities.

Multimedia Enabling of RDBMSs

Because of their great popularity, vast user base, and ease of use developed over 20 years, the RDBMS model is so well entrenched in the industry that no one expects the average user to abandon it. What is happening is stepwise upgrading and rearchitecting of RDBMS products to support complex data types beginning with images and voice messages that are characterized by relatively small amounts of data.

RDBMS products are optimized to support transactional processing of relatively small data items and are not well suited for handling massive time-sensitive and continuous multimedia traffic. Nevertheless, RDBMS vendors are well aware that they will have to provide multimedia capabilities with their products if they want to remain competitive. Most are in various stages of developing such features, but their approaches vary widely.

Enabling efforts to provide multimedia capabilities for existing RDBMS systems range from developing an adjunct engine to actually rewriting the database engine itself over a period of time. The latter is the philosophy of Oracle, a market leader in the RDBMS business.

One approach includes the creation of a transparent interface between RDBMS systems and a specialized ODB product that can handle multimedia or BLOBs. This is similar to what hybrid RDBMS products offer, but it is not such a streamlined product and does not necessarily provide built-in multimedia manipulating capabilities or optimization for managing BLOBs and bandwidth requirements.

Another approach includes the development of an object-oriented interface on top of the existing RDBMS product. This philosophy was followed by Hewlett-Packard in development of their Open ODB product line, which is basically an object-oriented layer on top of the AllBase/SQL (IBM's Structured Query Language) RDBMS product.

Other vendors have made less ambitious efforts which include options to include object management extensions in form of upgrading kits, advanced BLOB support with limited data management functions, inclusion of object-oriented development tools, and incremental additions of object functionality.

In most cases these solutions are seen as temporary because RDBMS vendors will certainly develop full multimedia object handling capabilities as time progresses and as multimedia networking becomes more common. Certainly without efficient multimedia capable RDBMS systems it is hard to imagine the future of the interactive enterprise and the virtual corporation. It also means that there is not much of a future for the ODB vendors, who will either merge with RDBMS firms who use their technologies or go out of business because they are unlikely to capture an adequate market share for themselves to compete against multimedia-capable RDBMS products.

Hybrid Multimedia Database Systems

This form of database is a relatively new development and is basically a combination of existing RDBMS and ODB products into a hybrid system specifically designed to support a growing range of multimedia networked applications while retaining the unique advantages of the traditional RDBMS systems. This approach is believed to be more effective than adding multimedia object handling capabilities to existing RDBMS products.

Hybrid RDBMS incorporate a true RDBMS model with full support of the critical SQL language with additional capability for complex data types. Hybrid RDBMS in networked environments can share multimedia data among many users and manage the storage, retrieval, and updating of such files. These systems are also optimized to use bandwidth efficiently, which is a critical consideration in multimedia networking applications.

Object-oriented technology is nevertheless at the heart of hybrid RDBMS products. Their vendors, such as Montage Software and UniSQL, are hoping to capture a significant market share before traditional RDBMS suppliers provide more comprehensive multimedia support in their products. They are differentiating themselves from RDBMS products but stress the compatibility with existing legacy and RDBMS systems.

The hybrid RDBMS consists of an ODB and RDBMS linked with each other. They combine all the capabilities of RDBMS with object-oriented features and can use SQL and C languages. The multimedia capability results directly from the separate object-oriented features which allow the definition of images, maps, sounds, and methods of working with such objects. These include compression, filtering of colors, and ability to manipulate those objects from the RDBMS framework. Such features must be developed by the end user in RDBMS products with multimedia capabilities, and clearly the hybrid RDBMS offers a distinct advantage. What is not clear, however, is how long this situation will exist because most observers agree that eventually traditional RDBMS vendors will provide full multimedia capabilities with future versions of their products. The concept of a hybrid RDBMS is illustrated in Fig. 14.4.

Object Databases

Object databases (ODBs) meet most of the requirements for multimedia database applications and are considered ideal for this purpose. ODBs can handle any kind of digitized information and are flexible when structuring and interrelating objects. Basically ODBs offer traditional RDBMS functionality for objects which contain encapsulated information about their attributes and behavior. However, most ODBs require high-performance workstation platforms and run mostly under different UNIX operating systems, although a few also support OS/2, Windows 3.x, and Windows NT. ODBs are believed to have their greatest potential as systems for controlling access to conventional and multimedia data stored in various systems within and without an organization.

ODBs allow objects to be stored, retrieved, and shared in much the same way as data is stored in RDBMSs. ODBs offer a better way to store objects because they provide all the traditional database services without the overhead of dis-

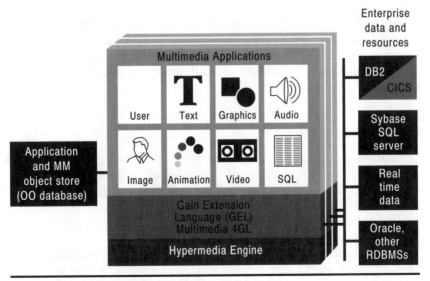

Figure 14.4 Hybrid RDBMS systems. (*Source: Sybase.*)

assembling and reassembling objects whenever they are stored and retrieved. As a result, ODBs are ideal as intelligent storage vehicles for multimedia data types, but this solution often requires a change in the mindset of developers used to dealing with conventional RDBMS products.

Within an ODB all data types are treated as objects and each has a relation to other objects, attributes, and records. These ODBs also come with built-in object-oriented tools but require additional programming knowledge. About a dozen ODB products are on the market, but the technology is relatively new and ODB vendors are mostly small ventures whose future is still uncertain.

The strengths of ODBs are in the fact that they support complex data types and BLOBs including composite objects. This is more representative of real-world information which is a more natural, intuitive way of structuring a database. Objects are self-contained modular pieces of information that lend themselves to being distributed around networks and servers which provide great flexibility to multimedia databases of this category. ODBs are the best way to deal with multimedia data types and BLOBs because these are more flexible than RDBMS solutions. ODB based systems permit addition of subclasses of objects without restructuring and manipulation of complex data quickly using direct links between objects, thus eliminating the need for table searches. System changes can be made with minimum coding.

Multimedia databases based on RDBMS solutions are inferior to ODB solutions because complex multimedia data types or BLOBs are inconsistent with RDBMS philosophy. Although some vendors began to add BLOB support to their products, the RDBMS model is based on tables of rows and columns of structured alphanumeric data. As a result, BLOB support in RDBMS products is at a primitive level confined primarily to storage and retrieval of BLOBs.

However, there are also limitations inherent in ODB-based multimedia databases. These include the need to provide object directories that know where the objects are stored, who is using them, and who is authorized to obtain which objects. In addition, ODB is a new and still immature technology which carries early adopter risks. These include lack of standards, difficulty in mixing objects from different vendor ODBs, unfamiliarity with object-oriented programming tools, and vendor instability.

Hierarchical Multimedia Databases

Multimedia has a very significant impact on storage requirements, particularly when large quantities of digitized video data are involved. It is also expected that analog storage devices such as audiotape recorders, VCRs, laserdisk players, and videocameras will remain in use as input devices for multimedia applications for a long time to come and must be integrated into multimedia networks at various levels.

In networked multimedia applications such as interactive TV, corporate broadcasting, and collaborative multimedia conferencing there is a need for massive multimedia storage hierarchies which will include magnetic, optical, and analog devices and systems. Such massive multimedia databases based on heterogeneous storage environments require complex data management systems which combine the capabilities of RDBMS and ODB with specialized physical file management systems. Figure 14.5 illustrates the concept of a hierarchical multimedia database storage system.

Multimedia Network Servers

Interactive multimedia computing is heavily dependent on client-server technology, permitting information storage and distribution at strategic LAN locations rather than centralized on a single mainframe. This approach allows PCs and midrange and mainframe computers to interact with each other on the network. On the *client* side of the client-server system it provides all facilities relating to the user, including data and query formats and how they appear on the screen. On the *server* side the system includes all functions that relate to the management and maintenance of data.

Interactive multimedia communications on an enterprisewide basis requires massive transmissions of audio and video datastreams that can rapidly take over all the available bandwidth of a LAN. Morover, such operations will interfere with normal data communications traffic, and strain the capabilities of existing client-servers beyond their capacities.

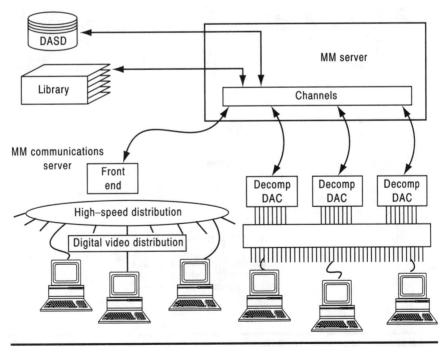

Figure 14.5 Hierarchical multimedia database storage system. (*Source: IBM Corporation.*)

Even without multimedia traffic, a number of high-performance dedicated client-servers can quickly saturate the bandwidth of a LAN. One solution is to link client-servers to a fast backbone like an FDDI network and connect LAN users to it directly or through a specialized hub as illustrated in Fig. 14.6. This type of approach may ease the congestion due to multimedia traffic, but some networking protocols may not be able to easily accommodate multimedia requirements to control very large datastreams with audio and video content. Other solutions which seem to be gaining acceptance include dedicated multimedia videoservers that store and transmit video data exclusively without interfering with existing traffic or applications on the LAN.

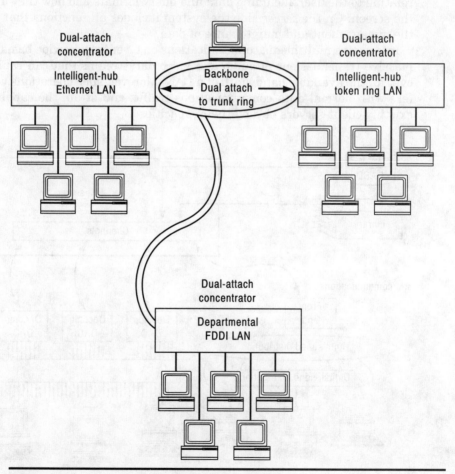

Figure 14.6 Multimedia traffic on a high-speed backbone.

Specialized Videoservers

A *videoserver* is basically a very large storage system that can operate as a virtual VCR for a large number of simultaneous users. It is a new concept which is still being developed, but major computer hardware manufacturers and some new ventures are already rushing in with solutions to what they see is a key multimedia networking market segment.

The videoserver is considered to be a crucial concept and a key element in two rapidly growing multimedia market segments. These include the collaborative multimedia of the interactive corporation and the consumer interactive TV market segments. Although videoserver hardware will vary in size and scope in those markets, the concepts remain the same and each will undoubtedly learn from the other as time progresses. More importantly, the existence of huge markets on both the corporate and consumer sides creates an incentive for vendors to bring new products to market at very competitive prices as soon as possible.

In the corporate world the videoserver is key to support of collaborative multimedia activities at the individual, group, and enterprisewide levels leading to the interactive enterprise status. This market also overlaps and often incorporates desktop videoconferencing and corporate video broadcasting, and in the future these functions are expected to merge. Videoservers will also play a major role in any customer services, interactive advertising, and kiosk merchandising, whether delivered via private networks or available through access to the digital superhighway in the future.

The consumer market segment includes the much larger back-end videoservers to support interactive TV and other services targeted at the individual households. It is expected that by the end of the 1990s the videoserver market will explode from practically nothing in 1993 to about $20 billion. This expectation of rapid growth is an incentive to computer manufacturers to enter this market as soon as possible. A major incentive to do this are several interactive TV pilot projects that are now getting under way and intensifying competition among cable, broadcast, and telephone companies.

Most computer manufacturers are using off-the-shelf high-performance hardware in assembling cost-effective solutions that can store and transmit huge volumes of video data with tolerable user response times. They face here the same problem of network latency in moving synchronized audio and video across complex networks, which must be kept at about 100 ms in order to avoid detection by the end users. Table 14.2 compares specialized videoservers from several suppliers.

Anatomy of a Videoserver

A multimedia videoserver is designed to store, record, and retrieve time-sensitive multimedia traffic such as video and audio, and interactive multimedia within existing client-server networks. Arguments abound whether such a sys-

TABLE 14.2   Typical Multimedia Videoservers

Product	Videoserver description
Mediaserver	Starlight Networks, Mountain View, CA Based on 486 50 MHz cpu, 16 Mbytes/RAM Server storage capacity to 100 Mbytes 10Base-T and thin-wire Ethernet LANs Token ring and FDDI support Up to 40 simultaneous users Media Transport Protocol with start/stop controls Video session services include stream management, object management, and video file management Application services include authoring tools Allows video applications to run with existing network equipment and cabling
Mediaserver	Oracle, Redwood Shores, CA Extension to Oracle 7 database Runs on massively parallel computer such as nCube Can deliver up to 25,000 discrete datastreams simultaneously and will increase tenfold Uses SQL Net software Designed initially for video-on-demand applications
Fluentlinks	Novell Multimedia, Natick, MA (previously Fluent Machines) NetWare Loadable Module for NetWare servers Ethernet and FDDI LANs Up to 40 simultaneous users Fluentview—video playback capability Fluentpresent—software only video playback/edit Fluentcreate—hardware assist video capture, edit and playback
Videocomm	Protocomm Inc., Trevose, PA NetWare Loadable Module Compatible with PCs and Macintosh Ethernet, token ring, and FDDI LANs Up to 25 simultaneous users

tem videoserver functions should not be based on conventional client-servers with multimedia capabilities or on separate, specialized systems dedicated to handling multimedia data types. Traditional client-servers interface with existing network management systems, but the use of specialized multimedia videoservers introduces new protocols and layers of operating software that may introduce additional unwanted overhead and unexpected delays. Figure 14.7 illustrates the role of a multimedia videoserver in a network application.

A videoserver introduces a number of functions into the network that facilitate multimedia processing. It frames compressed video signals for transmission across a LAN with special DSP compression chips for efficient delivery of variable-bandwidth datastreams. Other functions of a videoserver include compression of analog TV and VHS video signals and establishing connectivity with circuit-switching services and public videoconferencing circuits. Videoservers can set up and manage multiparty desktop conferences using composite displays and store video clips for videomail, training, and other mul-

Starlight Networks developed its own video protocol to control multimedia traffic on LANs. Its Mediaserver and Starworks software lets users send start and stop messages even during the transmission of very long datastreams.

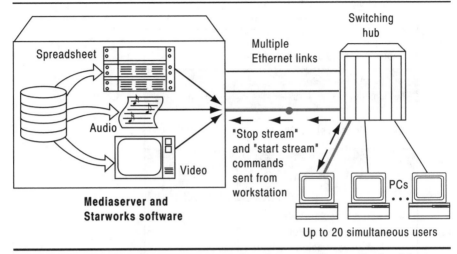

Figure 14.7   Multimedia videoserver in a network environment. (*Source: Reprinted from* Data Communications, *January 1993, p. 75, copyright by McGraw-Hill, Inc., all rights reserved.*)

timedia applications. All such functions could be developed in a conventional client-server, but the videoserver is designed to provide them transparently without additional effort. Videoservers may be based on specialized or customized hardware and massive storage including RAID or optical media systems. Figure 14.8 illustrates the different functions performed by a dedicated multimedia server.

Use of RAID Massive Storage Systems

Instead of conventional hard drives used in client-servers, multimedia servers improve their performance if used with special RAID multidrive arrays. In these systems instead of sequential storage, files are "striped" across multiple drives in such a way that they act as a single storage unit. These arrays offer performance enhancement, which increases linearly with the number of disks, but there is a practical limit of five-drive array which can deliver about 16 Mbytes per second of data.

Major Vendors

There are numerous vendors of multimedia file managers, some of which provide products that offer multiuser capabilities and networking support. These include Cato Software, Graphic Detail, IBM, Interactive Media, Lenel

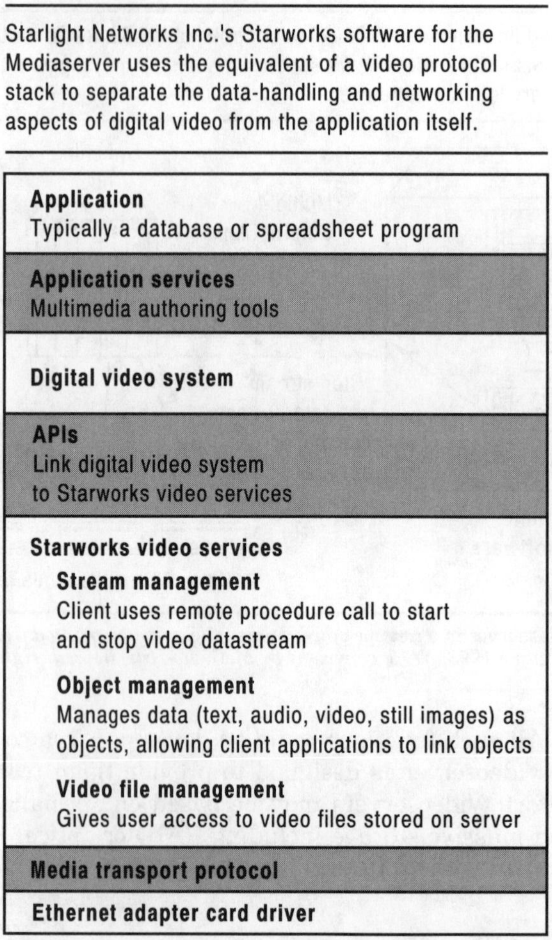

Starlight Networks Inc.'s Starworks software for the
Mediaserver uses the equivalent of a video protocol
stack to separate the data-handling and networking
aspects of digital video from the application itself.

Application Typically a database or spreadsheet program
Application services Multimedia authoring tools
Digital video system
APIs Link digital video system to Starworks video services
Starworks video services **Stream management** Client uses remote procedure call to start and stop video datastream **Object management** Manages data (text, audio, video, still images) as objects, allowing client applications to link objects **Video file management** Gives user access to video files stored on server
Media transport protocol
Ethernet adapter card driver

Figure 14.8 Multimedia server functions. (*Source:
Reprinted from* Data Communications, *October 1992,
p. 37, copyright by McGraw-Hill, Inc., all rights reserved.*)

Systems, Lotus Development, Multi-add Services, Northpoint Software,
Software Publishing, and Videomail.

Major vendors involved in multimedia capable RDBMS products include
ADB, The ASK Group, Borland International, Empress Software, Hewlett-
Packard, IBM, Informix Software, Montage Software, Objectivity, Oracle,
Sybase, UniSQL, and Versant Object Technology.

The specialized ODB vendors include ADB, Inc., Itasca Systems, Object
Design, Objectivity, Ontos, Servio, and Versant Object Technology.

Among specialized videoserver suppliers are companies like IBM, Novell
Multimedia Products (formerly Fluent Machines), Protocomm, and Starlight
Networks. Hewlett-Packard and Silicon Graphics are also involved in building
special videoserver systems for video-on-demand applications for the consumer
markets.

15

Multimedia Transmission Facilities

The major issue which impacts the future of collaborative multimedia communications, the interactive enterprise, and the virtual corporation is the universal existence and access to high-speed digital transmission facilities for reliable and cost-effective internetworking. The major issue that confronts the business that intends to get involved is the connectivity to these services. The proverbial last-mile connection is the most challenging and costly to implement. On the other hand, new services are being developed every day, prices are declining, usage is increasing, and governments are deregulating their telecommunication monopolies.

Current WAN alternatives include analog private and dial-up lines, digital private and dial-up services, frame relay, Switched Multimegabit Data Services (SMDS), and ATM. Not all options are available everywhere at an acceptable cost, and alternatives, if they exist, must be evaluated according to a number of criteria. These include the application requirements which in the case of interactive multimedia communications can be stated quite explicitly. It is the availability of a service to enable real-time interactive multipoint multimedia conferencing which defines bandwidth requirements depending on the number of simultaneous participants. For practical collaborative reasons it is probably no more than five to seven persons because groups with greater numbers of participants become unproductive and can be more effectively handled in a broadcast mode.

Specific criteria for evaluating a particular data transmission service must take into account the following factors:

- Maturity of standards in use
- Obsolescence of system in terms of buy or lease decisions
- Frequency of transmissions of multimedia traffic

- Availability of service in geographic locations
- Bandwidth capacities offered
- Delay tolerance that is acceptable
- Length of service anticipated
- Expenditures for equipment and services.

The situation in the telecommunications industry, which has been in constant flux recently, became even more volatile with the announcement and government sponsorship of the digital superhighway concept, formally known as the *National Information Infrastructure* (NII).

This chapter is primarily a general overview of the type and nature of telecommunication services that can provide some or all internetworking connectivity for multimedia communications. It is designed as on orientation in this vast and complex subject which merits a book of its own and which is covered much more exhaustively by numerous specialized publications and announcements from telecom carriers. Because services and their availability differ from one place to another, it is impossible to generalize, and the best policy is to contact local service providers to ascertain what specific services are available and at what cost.

National Information Infrastructure

The National Information Infrastructure (NII), which is popularly known as the *information superhighway,* assumes as its basic premise that there will be a digital line extended to every household in the country capable of handling interactive multimedia traffic. Before that happens, the objective, as voiced by Vice President Gore, is to provide a digital link to every classroom, library, clinic, and hospital and then extend it to the home.

According to executives responding to a survey developed by American Electronics Association, increased business efficiency is the most important reason for building the NII. Increasing competitiveness and facilitating research were given as the next two most important reasons.

The NII as a concept is a broadband interactive multimedia network serving business, education, and consumers. The Clinton Administration announced in its policy paper on NII that the government stands ready to change existing telecommunications laws, mandate new standards, and spend over $1 billion annually to make sure that the United States develops a high-capacity seamless web of wireless and wire-line communications.

Most of those funds are spent under the High-Performance Computing and Communications (HPCC) program, which finances such agencies as DARPA, National Science Foundation, Department of Energy, NASA, National Institutes of Health, NSA, NOAA, EPA (National Security Agency, National Oceanic and Atmospheric Administration, Environmental Protection Agency), and others. The 1995 budget proposes $1.3 billion to be spent on NII, representing a clear 32 percent increase over the budget of the previous year.

As an initial cooperative program with the business world, DoD granted $70.7 million to a consortium led by IBM to develop National Industrial Information Infrastructure Protocols, which will compile a definition of all products that facilitate the exchange of information.

In late 1993 a consortium of 20 organizations decided to construct an NII testbed. It includes implementation of the latest digital WAN technology using ATM and frame relay to link several research centers on east and west coasts with Sandia National Laboratories in Livermore, California, providing access to 20 years of data about ocean and land pollution. Aside from several universities, industry participants include AT&T, DEC, Essential Communications, Hewlett-Packard, Network Systems, Novell, Sun Microsystems, Sprint, and SynOptics Communications.

In December 1993 a number of leading companies formed a Cross-Industry Working Team designed to build consensus on the network services and applications that would be used by business, schools, and homes under the NII concept. The group will define technical requirements and provide feedback to all parties involved. Its major goal is to ensure standardization and promote interoperability. The group operates through a set of subgroups on architecture, applications, portability, and services. Primary members of the group include Apple, AT&T, Bellcore, Bell South Telecommunications, Cable TV Laboratories, DEC, GTE Laboratories, Hewlett-Packard, IBM, Intel, McCaw Cellular, MCI Communications, Motorola, NYNEX, Pacific Bell, Silicon Graphics, Southwestern Bell, and Sun Microsystems. Associate members include Cisco Systems, CBEMA, Financial Services Consortium, Hughes Network Systems, SAIC, Sprint, 3Com, West Publishing, and Xerox.

It will be some time before NII becomes a reality, but elements that will make up the network eventually are emerging piecemeal here and there. Mobile users with powerful laptops and notebooks are traveling throughout the country, and it is logical to expect that they might want to hook up and participate in multimedia conferencing wherever they may be at the time. This is only part of the pressure building up to create universal access, but there are other business considerations that are acting to retard that process.

Carriers are building the backbones based mostly on fiberoptic media to provide high-speed services to anyone who needs them but it's not much use to the person in the street or the home as long as the last mile remains a copper wire with minimal bandwidth capacity. But carriers are also reaping excellent revenues from existing digital links, and it is naive to expect that they will develop low-cost universal digital networks that will eliminate the need for their much more costly services.

As far as international information superhighways are concerned, a number of issues influence the developments and in fact slow down any progress. Politics of other countries are the main controlling forces. In major European countries where national monopolies operate telecommunications, there are major problems with employment in those companies. These organizations are among the largest employers in their countries and cannot allow introduction of advanced technologies that will drastically reduce the workforces in their industry. As a result, a global information superhighway is generally seen to be some years away. What this means to the interactive enterprise and virtual corporation is reliance on the more costly and not so ubiquitous private and special digital services developed by the independent carriers.

Wide Area Transmission Services

A number of alternative wide area transmission services are in existence that provide connectivity between PCs and LANs. These include private data networks, public data services, leased circuits, ISDN services, or analog telephone lines. Each of those alternatives provides a variety of bandwidth capacities, but most fall short of supplying a cost-effective facility for collaborative multimedia communications. Nevertheless, these services are being constantly upgraded, and new offerings with some multimedia capabilities are being introduced. For this reason it is important to keep in mind all the available alternatives and tailor specific multimedia applications to take advantage of a particular service or services that present the most cost-effective solution. Table 15.1 presents an overview comparing the different services available and their respective bandwidth capacities.

Analog Telephone Networks

Plain Old Telephone Service (POTS) is a service provided over copper wires in analog mode and requires the use of a modem to transmit digital data. It represents the lowest bandwidth available, but it is ubiquitous, and its installed base is larger than that of any other network under development.

Given all these limitations, POTS is nevertheless a dial-up network and offers a number of transmission standards ranging from the obsolete 1200 bps to attempts at providing as much as 57.6 Kbps of bandwidth capacity. Table 15.2 outlines all POTS lines standards available and in development.

Limited low-resolution multimedia transmission is possible with POTS. The leading-edge POTS transmission standard V.32bis modem provides 14.4-Kbps bandwidth and coupled with a 4:1 compression can theoretically produce a 50-Kbps throughput over a typical business telephone line. In practice data can be transmitted at only 20 to 30 Kbps at best. However, V.32 Turbo standard provides 19.2 Kbps, and more bleeding-edge V-Fast standards are operating at 24, 28.8 Kbps, and higher speeds. Even faster speeds of 38.4 Kbps have been available but with difficulty, and 57.6 Kbps is possible but cannot be guaranteed. At those speeds transmission over analog networks hits the medium barrier, and many analysts believe that without actual enhancement of the copper lines themselves higher data transfer rates are not possible.

TABLE 15.1   Wide Area Transmission Services

Service	Bandwidth ranges	Comments
Telephone	2.4 Kbps universal 9.6 Kbps attainable 19.2 Kbps not always available 38.4 Kbps difficult to obtain 57.6 Kbps no guarantee of	Low cost Very slow Universal connectivity Best for small transactions
Leased lines	T-1 ≤1.544 Mbps T-2 6.312 Mbps T-3 46 Mbps T-4 273 Mbps	Not designed for LANs Can be very expensive Not available in all locations
X.25 packet switching	9.6–56 Kbps	Relatively low cost High networking overhead Best for bulletin boards
Switched 56	56 Kbps	Moderate costs Single-channel service Best for data transmission Not available everywhere
ISDN	Basic rate 56–144 Kbps Primary rate up to 1.54 Mbps	Expensive to connect Low-speed applications Not available in all regions Telecommuting potential
Frame relay	56 Kbps to 1.544 Mbps	LAN-WAN potential Higher bandwidth possible Intermediate potential
SMDS	1.544-Mbps access 46-Mbps operation	High cost Intercompany transmissions Group networking Scalable bandwidth Can support LAN speeds May evolve to ATM
ATM	45 Mbps 150 Mbps 600 Mbps 2.48-Gbps potential	Broadband ISDN standard Very high speeds Expensive at present Universal access potential Virtual LAN potential

The AT&T VideoPhone 2500 color video communications across POTS uses V.32bis modems and provides a very small frame of jerky 10-fps images. Similar results are obtained with competitive products that can at best reach about 12 fps in a very small window which does not provide a sense of continuous motion but may suffice for simple point-to-point videotelephony. Needless to say, the price of POTS circuits with the high-speed transmission capacities increases from cents to hundreds and thousands of dollars.

TABLE 15.2   Analog Telephone Network Standards

Standards	Status	Speed, bps
212A	Obsolete	1,200
V.22bis	Mature	2,400
V.32	Standard	9,600
V.32bis	Leading edge	14,400
V.32 Turbo	Advanced concept	19,200
V-Fast(1)	Advanced concept	24,000
V-Fast(2)	Advanced concept	28,800
	Difficult to attain	38,400
	Possible unguaranteed	57,600

Satellite Communications

The very-small-aperture terminal (VSAT) satellite systems can provide transmission facilities ranging from 15 to 56 Kbps. Satellite communications are useful for connecting remote sites and can provide interactive services for conferencing facilities. The quality of the video in such transmissions may not be the best, and frame rates usually are relatively low. The delay factor in satellite transmission is too severe for useful collaborative multimedia traffic to be effective, but the service is adequate for transmitting images and video which is not time-sensitive in nature.

Satellites are also used for broadcast-quality transmission of programs to cable TV headend stations. Digital satellite TV systems use compression and lower the costs of cable TV channels. They are also prototypes of direct home satellite transmission with MPEG compression chips being installed in digital TV receiver sets. Numerous services providing direct digital TV channels are being developed and may provide part of the future interactive TV infrastructure.

Switched 56 Services

This is a network of relatively old single-channel leased copper lines with a very large installed base. Their bandwidth, which is 56 Kbps, makes them a better solution for videotelephony or even videoconferencing than POTS networks.

On the other hand, advanced modems can bring the same bandwidth to less expensive dial-up networks, and switched 56 does not offer multiple channels, which makes it impractical for multimedia conferencing. However, it is possible to obtain a $1/4$-screen video image playing at 10 to 15 fps on this bandwidth.

Switched 56 service offers certain advantages such as broadcasting connectivity, faster basic speed than analog lines, and usage-sensitive billing, which can result in lower operating costs. Major applications include LAN internetworking with dial-on-demand routing, videoconferencing, digital audio transmissions, backup of dedicated lines, and Group IV facsimile transmission.

Integrated Services Digital Network

Integrated Services Digital Network (ISDN) is the integration of communication services over digital facilities, including wire, coaxial cable, optical fibers, microwave radio, and satellite. It provides end-to-end digital connectivity between two communication devices and is capable of transporting voice, data, graphics, text, and video over the same lines. Figure 15.1 illustrates the basic ISDN service concept. ISDN is available in two distinct service offerings as a *basic-rate* ISDN with two channels and as *primary-rate* ISDN which includes 23 channels.

ISDN is now the most common communications service used by traditional videoconferencing systems. The bandwidth capacity provides the basic videoconferencing video, but with the high compression involved in these systems, the image will not be stable if people move quickly in front of the camera, although the audio quality is not affected. The multiple channels permit simultaneous data and fax transmissions, which is a valuable feature in workgroup collaboration.

ISDN requires a digital switch, and although RBOCs are upgrading their facilities, ISDN is not available universally but rather mostly in campus and large company environments. A single access connects the user to the entire range of public communication networks for voice, image, text, data, and video

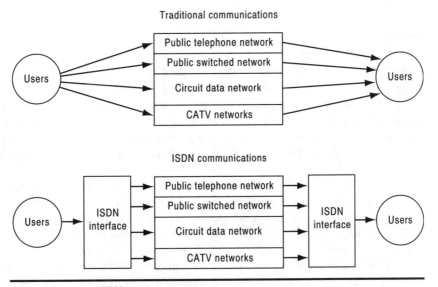

Figure 15.1 Basic ISDN service concept.

by circuit-switched or packet-mode transmissions. A national ISDN network which will connect central office switches with one another is being implemented but it is not expected to be complete until the end of the century.

A major problem with ISDN services is access. Geography, government regulations, and competitive pressures have inhibited the spread of ISDN services even though scores of companies are in the market for exactly that type of service. The biggest issue is availability of digital switching in the local loop also known as the proverbial last mile, which in fact may often run to 3.5 mi and is predominantly analog in nature. ISDN cannot become a national standard until all local phone switches and loops are upgraded. This is further complicated by the fact that digital ISDN signal is reliable only 18,000 ft from the switch. This means that even if a digital switch already exists in a local switching office, the end user must be located within 18,000 ft of the switch to reliably use ISDN services.

Basic-rate ISDN

Basic-rate ISDN offers a total bandwidth of 144 Kbps which includes two 56- or 64-Kbps channels that carry voice and data and a 16-Kbps data link for signal control. The two 64-Kbps channels may be combined to provide a 128-Kbps line if necessary. ISDN is also less expensive than leased lines, with prices of basic-rate ISDN in the United States comparable to POTS service.

Basic-rate ISDN is now being marketed as a service to small businesses and telecommuters, but its acceptance is slow because of relatively high prices. Nevertheless, it is replacing leased circuits for dial-up conferencing and is increasingly used to transmit multimedia traffic both in the United States and in Europe.

Primary-rate ISDN

Primary-rate ISDN is basically an interface with 1.544-Mbps bandwidth, which is the same bit rate as that of the T-1 lines. It consists of 24×64-Kbps channels, one of which operates as a signaling channel and packet data channel. This interface is flexible and allows combinations of several 64-Kbps channels to provide higher-speed services such as 384 Kbps. Additional developments on this service will provide more bandwidth choices for the end user in the lower bandwidth range of 8, 16, and 32 Kbps below the basic 64-Kbps channel speed.

ISDN is in fact designed to provide bandwidth on demand, and the user can specify the data rate necessary on a call-by-call basis. Another key component of ISDN is performance on demand since that factor also varies widely from one application to another. It is seen as a useful service in providing the backbones for internetworking between LANs and is already widely available for this purpose in Europe and Japan. As a digital backbone ISDN offers network-based fault tolerance because if a link or regional node fails, users can still access the backbone by dialing another regional node. On the other hand, primary-rate ISDN requires relatively expensive equipment to obtain access to the service.

Some forward-looking companies made the effort to adopt ISDN as a WAN backbone while waiting for more universal access to become available. In some cases companies like Moore Data Management Services joined the local carrier US West in a partnership to test and develop their ISDN service. Nevertheless, companies have recently become more interested in linking LANs at their native speeds of 10 and 16 Mbps, which the ISDN services cannot deliver. This puts a question mark on the future of ISDN, particularly in view of the development of broadband ISDN (BISDN) services that provide such capabilities in form ATM and SMDS services and are being offered by the same carriers.

Frame Relay

Frame relay is a bridge between existing narrowband and future high-speed broadband transmission services with a potential to provide bandwidth capacity over 2 Mbps and an easy migration path to ATM.

Frame relay, like packet switching, combines data into frames and allocates bandwidth to multiple datastreams. Unlike the X.25 packet-switching standard, frame relay does not involve error correction, which makes it faster and more efficient. The digital hardware facilities of today are apparently so much more reliable that error-correction procedures of the past are redundant and time-consuming.

Frame relay is not suitable for transmission of voice or video because it does not provide constant latency transmission and is essentially designed for high transmission speeds between LANs. However, it is seen also as an interim form of service as migration strategies to higher-speed services such as ATM are being developed. A typical frame relay network is illustrated in Fig. 15.2.

Two types of frame relay services exist. One is a premium service which is national and global in scope, while the other provides local and regional facilities. Frame relay services offer a range of speeds which include 56/64, 112/128, 256, 384, 512, 768 Kbps, and 1.026 and 1.536 Mbps. Interexchange carriers offer the whole range of speeds, but the local services usually provide only 56 Kbps, 384 Kbps, and T-1 or 1.536 Mbps.

Basically customers use dedicated T-1 or 56-Kbps lines to reach interexchange frame relay services, but most carriers are planning to offer high-speed dial-up access as well as through ISDN circuits. Nevertheless, frame relay implies switches, routers, bridges, and interfaces, all of which have a cumulative latency effect on any time-sensitive traffic.

There are already 16 domestic frame relay carriers, of which 7 provide international services and another 6 are foreign carriers. The need for low-cost international connectivity is driving the frame relay market which

A frame relay network consists of backbone switches and a variety of access devices, including routers, concentrators, and frame relay access devices (FRADs). Frame relay interfaces also can be added to workstations in the form of plug-in cards and software.

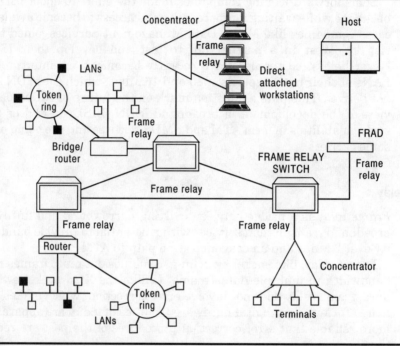

Figure 15.2 A typical frame relay network. (*Source: Reprinted from* Data Communications, *May 1992, p. 70, copyright by McGraw-Hill, Inc., all rights reserved.*)

already exists in 25 countries in which one or more carrier provides frame relay services.

Major premium frame relay service providers include AT&T, BT plc, Cable & Wireless, CompuServe, MCI Communications, PacNet, Sprint, and WilTel. Regional carriers include Ameritech, Bell Atlantic, Bell South Telecommunications, EMI Communications, NYNEX, Pacific Bell, Southwestern Bell, and US West Communications (Table 15.3).

According to a forecast from Vertical Systems Group of Dedham, Massachusetts, the worldwide frame relay market is expected to grow from under $100 million in 1993 to almost $1 billion by 1996.

High-Speed Leased Lines

Transmission facilities with higher bandwidth than frame relay and primary-rate ISDN are provided by circuit-switched trunk lines based on time-division multiplexing (TDM) technology. Circuit switching transmits data bits or other digital traffic such as digitized voice or video transparently between two end-

TABLE 15.3   Frame Relay Services

Service provider	Type of service	Comments
Ameritech	Premium service	Supports network-to-network standards
AT&T	Premium service	Supports NNI standards, 17 countries
Bell Atlantic	Regional	
Bell South	Regional	
BT North America	Four speeds	
Cable & Wireless	Premium service	
CompuServe	Premium service	Nine countries
EMI Communications	Regional	
MCI Communications	Premium service	Twelve countries, 391 POPs
NYNEX	Regional	Supports NNI standards
Pacific Bell	Regional	
PacNET	Premium service	
Southwestern Bell	Regional	Supports NNI standards
Sprint	Premium service	Fourteen countries, 330 POPs, supports NNI standards, usage-based pricing
US West	Regional	Supports NNI standards, 75 POPs
Wiltel	Premium service	First commercial service, supports NNI standards, intends to lead in ATM backbone services

points of a call. Since data is not interpreted en route, circuit switching is protocol-independent and can transmit large amounts of bandwidth with very low delay, and virtually any traffic can be supported by those lines.

This technology originated in the public voice networks and has been used on trunk lines between central offices since the 1950s. Until the early 1980s carriers and governments were the exclusive users of those trunks, but deregulation and competition dramatically reduced the cost of high-speed circuits and made them available to end users. Since then carriers introduced more sophisticated leased-line services which can be accessed with appropriate interface equipment.

These trunk facilities are available as T-1, T-2, T-3, and T-4 lines which provide bandwidth capacities ranging from 1.544 Mbps all the way to 273 Mbps (Table 15.4).

Fractional T-1 and T-1

The next-highest bandwidth service above frame relay is the *fractional T-1,* sometimes known as *switched 384,* which transmits at 384 Kbps, a rate that uses only a fraction of the 1.544 Mbps. This fractional T-1 also happens to be the barest minimum for transmission and display of a full-screen, full-motion

TABLE 15.4   Bandwidth Levels of Leased Trunk Lines

Trunk line	Digital signal level	Speed, Mbps
T-1	DS-1	1.53
T-2	DS-2	6.31
T-3	DS-3	44.70
T-4	DS-4	274.00

video at 30 fps (frames per second). As such, fractional T-1 can support good-quality small-group videoconferencing sessions with use of appropriate codecs.

Fractional T-1, which was introduced in the late 1980s, enables even small users to take advantage of T-1 pricing and allows use of single circuits of 56 or 64 Kbps as required. Fractional T-1 provides transport for a set of bit rates in multiples of 56 or 64 Kbps.

The T-1 facilities are the medium of choice for high-quality videoconferencing and are commonly used by MIS departments for linking computers and LANs. It was developed to provide 24 voice channels, each with 64-Kbps bandwidth and an additional 8 Kbps for synchronization and framing, which brings the total up to 1.544 Mbps.

The T-1 lines provide the same 1.544-Mbps bandwidth as primary-rate ISDN but are much more widely available, particularly in the United States, Japan, and Korea. A comparable service with a bandwidth of 2.048 Mbps known as *E-1* is available in some parts of Europe, Australia, Asia, the Middle East, and South America.

T-2 leased lines

This line speed is based on the DS-2-level standard used to transmit formatted digital signals at 6.312 Mbps. It is equivalent to 96 voice channels on a single line.

T-3 leased lines

The T-3 line offers the equivalent bandwidth of about 28 T-1 lines or 45-Mbps capacity, and T-3 is the main distribution backbone of the telephone networks. T-3 can provide broadcast TV-quality video transmission, but it is used mainly for very data-intensive applications such as real-time color 3-D visualization modeling. T-3 bandwidth is comparable to SMDS, which can carry voice, video, and data at 45 Mbps, but it is much more widely available.

T-4 leased lines

The T-4 is the highest-level leased-line facility, with a bandwidth of 273 Mbps, which is equivalent to 4032 voice channels.

Switched Multimegabit Data Services (SMDS)

SMDS is a newly created cell-switched data transmission service designed for interconnecting LANs, computers, and workstations over WAN with LAN-like performance. It is best suited for users whose applications require high bandwidth for part of the time such as transmission of detailed images or maps which are voluminous but not time-sensitive.

SMDS is a connectionless cell-switched service originated by Bellcore with the objective of developing a high-speed, ubiquitous switched data service based on the cell relay technique which in effect is a fast and efficient form of packet switching. SMDS does not require carrier switches to establish a path for data transmission. Basically SMDS slices up data packets up to 9 kbytes in size into a set of 53-byte cells which are switched through the carrier network over any available path to the destination where they are reassembled.

The 53-byte cell size was purposely set by Bellcore to align SMDS with ATM. In both cases the cells are compatible with regard to size, address, and control headings, making for easy migration from SMDS to higher-speed services. This also suggests that SMDS is a temporary service with a relatively limited future and only local appeal.

SMDS is comparable to a high-speed POTS. If someone's SMDS address is known, data can be exchanged immediately, but generally SMDS services are available within a local access and transport area (LATA) of a particular regional telephone company. Long-distance SMDS services can be made available through interconnections by MCI Telecommunications. Table 15.5 outlines SMDS service coverage by region at the outset of 1994.

SMDS offers a choice of speeds including 1.17 Mbps with access through T-1 circuits and 4, 10, 16, 25, and 34 Mbps with access through T-3 circuits. Lower speeds, on the order of 54 and 64 Kbps, are being made available through a special interface. Connectivity is established through an interface card and channel service unit which carry a one-time installation charge, but otherwise SMDS is thus far a fixed-price service, which means no mileage or usage

TABLE 15.5   Initial Availability of SMDS Services

Service provider	Regional availability
Ameritech	Ohio, Illinois, Michigan, Wisconsin
Bell Atlantic	Delaware, District of Columbia, Maryland, Pennsylvania, New Jersey, Virginia, West Virginia
Bell South	Atlanta, GA; Birmingham, AL; Charlotte, NC; Huntsville, AL; Nashville, TN; Miami, FL
GTE Telephone Operations	Dallas, TX; Everett, WA; Los Angeles, CA; Tampa, FL
Pacific Telesis	Anaheim, CA; Chico, CA; Fresno, CA; Los Angeles, CA
US West	Minneapolis, MN; Phoenix, AZ; Portland, OR; Salt Lake City, UT; Seattle, WA

charges. This is advantageous where distances between sites and usage are high.

The advantages of SMDS include scalability beyond T-1 speeds, ease of introduction of new users, bandwidth on demand, connectionless service, multicasting support for some protocols, and fixed pricing. Disadvantages include local (limited-distance) orientation, unsuitability for time-sensitive transmissions, interim character as a step toward ATM, and relatively brief history with only a few hundred users. This is reflected in the projections for SMDS service revenues which are forecast at only $85 million in 1995, rising to $220 million in 1997.

Broadband ISDN

Broadband ISDN (B-ISDN) is a high-speed wide area telecommunications service which was designed to use ATM as its switching methodology. Since then ATM has taken on a life of its own, and other high-speed services such as SMDS and SONET also have roots in B-ISDN. As a result, B-ISDN can be seen as a basis for a set of several high-speed services whose speeds begin at 150 Mbps. The 150-Mbps speed was chosen as a starting point because it is the slowest speed defined by SDH, the international SONET networking standard. The ATM switching methodology, which incidentally was originally designed for SONET, is discussed in detail in Chap. 14.

In the context of B-ISDN, ATM can be seen as a connection-oriented B-ISDN mode, while SDMS a connectionless version. SONET, by the same token, can be defined as an international high-speed multiplexed transmission standard for fiberoptic devices and interfaces that are used on B-ISDN services.

Basically B-ISDN is a new switching and multiplexing concept which specifies the use of fixed-length packets known as *cells*. These cells are specifically only 53 bytes in length and do not provide for detection of errors in user data. Such simplifications of the packet-switching technique allow for data throughput increases of one or two orders of magnitude in B-ISDN networks. The use of fixed-length cells also allows introduction of new hardware switching techniques as faster microchips and hardware become available.

B-ISDN overcomes some of the limitations inherent in frame relay services. The small size of the cell makes it very suitable for handling time-sensitive traffic such as audio and video. The basic cell relay concept is representing a single unifying technology which is very flexible in handling a different mix of transmission speeds and delay tolerances. As a result, B-ISDN is meant to serve as a backbone technology in public networks. It is designed to support mostly T-3 and higher-speed transmission facilities. Although B-ISDN switches are beginning to appear in central offices, it will be some years before their widespread deployment will take place.

Synchronous Optical Network

The *Synchronous Optical Network* (SONET) is a high-speed multiplexed transmission standard for fiberoptic devices and interfaces used on B-ISDN networks. It was developed specifically to meet the needs for transmission speeds above the T-3 level of 46 Mbps and is designed primarily to reduce the congestion on WANs backbone networks. SONET specifies intermediate speeds from 51.84 through 2488 Mbps up to and including 4976 Mbps or almost 5 Gbps. The international version of SONET is known as *Synchronous Digital Hierarchy* (SDH) with speeds increasing in increments of 155 to 622 Mbps and 2488 Mbps.

Many believe that SONET is a step in the development of photonic switches which operate with terabit-per-second (Tbps) speeds and can handle 10-Gbps nodes. This type of technology is believed to be necessary to provide true telepresence during virtual meetings and conferences with 3-D representations of people and objects. Nevertheless, this type of multimedia communications is about a decade away from today. Adaptive Corporation of Redwood City, California, and T3plus Networking, Inc. of Santa Clara, California, are two vendors who are already offering SONET interfaces in their networking equipment.

The SONET standard is based on a fixed frame format which contains management, maintenance, and overhead information as well as user data. Such a format ensures that optical transmission equipment from different manufacturers is compatible.

Cell relay technology is the basis for SONET protocols applied to BISDN and ATM, both of which are basically packet-switching techniques using very small packets of fixed length. The objective of this approach is specifically to minimize processing and transmission latencies and obtain high transmission speeds. Telephone companies are using SONET for connections between central offices and exchanges, while cable TV firms are interested in eventually exploiting it to reach into the home.

Telecommunications Carriers Multimedia Strategies

The carriers are aware that the users are moving away from rigid private-line networks to the more dynamic digital services. However, the major motivation is alleviation of internetworking congestion between LANs and WANs rather than facilitation of interactive multimedia networking. As a result, the carriers are more concerned about meeting those client needs than promoting multimedia networking.

From a purely business standpoint, the carriers are also interested in exploiting their existing facilities such as the trunk lines, which are much more widely available than the new high-speed service offerings. Nevertheless, they see frame relay, SMDS, and ATM as three services that dovetail each other with regard to bandwidth offered and present a clear migration hierarchy toward high-speed networking which will eventually include universal collaborative multimedia operations.

The rather slow pace at which ISDN services have been introduced is another indication of reluctance to invest in transmission technologies that may be superseded or rendered obsolete before a return on investment is realized. The range of speeds offered on ISDN overlaps those available on more expensive leased lines; however, the proliferation of LANs and WANs during that period also impacted the carriers' reluctance to proceed more vigorously with providing those services.

The growing demand for efficient LAN-WAN connectivity is clearly conducive to the development of frame relay services which offer higher bandwidth that are necessary for internetworking. Unlike private leased lines, frame relay allows bursting above contracted capacity and is more dynamic and useful under the networking loads that are also growing rapidly and often unpredictably.

As far as SMDS is concerned, the ability to offer LAN speeds and connectivity make SMDS services attractive to many users. However, SMDS services are not offered by all the carriers, and some carriers are bypassing SMDS altogether, concentrating on development of ATM services. Since ATM offers a more flexible service, including bandwidth on demand and full multimedia capabilities, many users may be willing to wait, but competitive pressures will force them to use whatever high-speed services are available. In fact all carriers regard frame relay and SMDS as a temporary stopgap solution until ATM markets are clearly defined.

The carriers are also somewhat influenced by independent communications hardware vendors who are competing by bringing out new products, particularly in the high-speed internetworking that facilitate migration from frame relay to ATM. Some of these products allow users to combine their private leased lines and public packet services and provide means of managing the transmission process. Such facilities make it possible for users to buy different services at different times of day and for different applications by automatically optimizing the cost of such services. The carriers are also aware that as time progresses specialized multimedia transmission services in specific industry application areas will come into play. This will relieve the carriers from taking unknown risks and give them a free hand to concentrate on providing and maintaining high-speed transmission facilities.

16

Multimedia
Networking Hardware

Multimedia networking is becoming possible as a result of the convergence of several technologies and activities, including high-performance PCs and workstations, specialized videoservers, multimedia databases, massive storage systems, high-speed LANs, high-bandwidth data transmission services, more efficient compression devices, multimedia operating and development software, and numerous new agreements on operating standards.

Most of these technologies have a life of their own and are taking place without regard to multimedia as such, and we have discussed the development of multimedia networking within these technologies and environments in previous chapters. It is true to say that those technologies facilitate the development of multimedia applications and the conceptualization, design, testing, and deployment of interactive multimedia communications in all its forms. On the other hand, many existing products that go to make up a LAN or an internetworking system of LANs and WANs are not multimedia-capable, and those systems would have to be upgraded to support interactive multimedia communications.

As a result, a whole new group of hardware products is emerging that are now specifically designed to perform all the conventional tasks and provide multimedia capabilities. Other devices are specifically designed to handle multimedia processing but also have wider networking and information processing implications. This chapter is an overview of such new multimedia networking hardware products in order to provide the reader with a vision of the future of multimedia networking and its business potential. It covers specific multimedia networking products which do or will enable collaborative multimedia networking and communications and that have not been covered in previous chapters.

Enabling Technologies

Traditional multimedia, which is basically centered around standalone PC-based multimedia platforms, consists of an assembly of various hardware components and associated software. These include audio and video digitization and compression boards, CD-ROM drives, videocameras, speakers, and microphones all attached to a sufficiently powerful microcomputer. Operating systems for these machines include multimedia extensions or more recently built-in capabilities to facilitate the operation of all these devices in a coordinated manner as multimedia platforms.

At least three dozen multimedia upgrading kits are on the market designed to convert conventional PCs into multimedia platforms with the minimum of effort and relatively little cost. The next generation of multimedia-capable PCs is already on its way with built-in audio and video compression microchips on the motherboard and internal CD-ROM drives as standard features.

Those who are interested in developing and distributing multimedia applications can draw on an additional set of enabling technologies in form of authoring systems, sound and video editing systems, image capturing and manipulating devices, and CD-R optical recording hardware and software products.

Basically all those are seen as enabling technologies of multimedia delivery and development platforms, and special standards have been adopted such as MPC-1 and MPC-2 to designate the required performance levels of various components that combine to provide an acceptable multimedia platform.

It is our belief that in the near future a large percentage of the existing and growing PC population will be enabled to process third-party multimedia applications and that some will also enhance their systems sufficiently to become multimedia application creators and developers.

From the perspective of multimedia networking, these are important developments because unless the basic individual PCs and workstations are enabled as multimedia platforms, the concept of collaborative multimedia operations and interactive enterprise cannot be realized.

Over and above the need to enable PCs for multimedia individually there is the separate and more complex requirement for enabling the LANs and WANs to handle multimedia-enabled PCs and workstations and provide the networking infrastructure to engage in interactive multimedia communications. What this means is a whole new set of enabling technologies which include DSPs, interfaces, gateways, bridges, routers, switches, hubs, and multipoint control devices.

Future Hardware Requirements

In order to understand where multimedia networking is today and what is needed to achieve the ultimate objective of real-time collaborative multimedia communications as defined by the virtual LAN and the interactive enterprise, it is instructive to take a look into the future. One of the most useful insights on this subject comes from research conducted by Texas Instruments recently. The company investigated this subject to ascertain what types of microchips should be designed and produced in the near future in order to achieve such multimedia objectives on a global basis.

Specialized semiconductor devices that can handle JPEG, MPEG, H.261, and many proprietary compression algorithms on a single chip are seen as the future building bricks of multimedia hardware products. TI already developed a general-purpose Multimedia Video Processor (MVP) along those lines which is specifically targeted at the "integrated media workstations" that will provide the platforms for handling real-time multimedia transmissions. Such microchips process input from cameras and video software, generate graphics and animation, and offer very high-resolution, full-motion video capabilities. The company believes that MVP-type microchips will be in great demand by interactive multimedia communications hardware developers particularly for use in hardware products to facilitate imaging and complex document transmissions.

According to TI research, full-fledged interactive multimedia transmissions under the H.261 standard on an international basis will require hardware products with performance characteristics in the order of 1 billion operations per second (BOPS). This is about 10 times as much as the most powerful workstations on the market today.

TI has already announced the development of a 2-BOPS MVP device that combines DSP and reduced instruction set computer (RISC) functions and integrates all video technologies on a single chip. Such devices operate at data transfer rates of 400 Mbps at the minimum to produce smooth, real-time multimedia transmissions and will employ 3 to 4 million transistors.

Current multimedia networking hardware is nowhere near these capabilities, so it is futile to expect ideal solutions for collaborative multimedia just yet. Nevertheless, it is useful to keep in mind those requirements and developments when planning multimedia networks because the gap between what is available now and what will come to market in only a few years will be closing very rapidly.

Developments in the semiconductor industry are among the most reliable indicators of the immediate future of multimedia hardware products. MIS managers who keep track of those developments are not likely to be caught by surprise. They will always perceive more clearly all the alternatives ahead of their competitors, which will help them to protect their investments without sacrificing technological leadership.

Digital Signal Processors

Because DSPs are so crucial to the development of multimedia capable hardware, it is worth looking at this basic component in more detail. DSPs rapidly translate analog signals of the real world into digital patterns and are now seen as crucial to making real-time interactive multimedia processing hardware practical and cost-effective. Originally DSPs were developed to handle radar and other mission-critical signal acquisition and analysis functions. As a result, they evolved into a class of microprocessors optimized for handling large amounts of real-time data generated by analog inputs. They form an ideal match to the functions required for processing efficiently multimedia data. Many ASIC (application-specific integrated-circuit) products that were previously developed as compression and decompression chips or special image processors for particular product lines are now being replaced my more efficient DSPs.

The basic multimedia DSPs can handle real-time operations in partnership with host processors and DSP-based audio/video codecs, specialized compression chips, and voice and music applications. They are superior for such tasks compared with general-purpose CPUs because they can handle high interrupt rates with low interrupt latencies. DSPs interface more efficiently with devices for converting analog-to-digital (A/D) and digital-to-analog (D/A) signals. They are also a better match for the vector-oriented nature of most algorithms than even the latest high-performance RISC microchips. It has been pointed out by experts that a 20-MIPS (million instructions per second) DSP in a real-time signal processing algorithm is equivalent to a 60-MIPS general-purpose processor.

The latest breakthrough in multimedia microchips are DSPs with algorithms stored in RAMs which enable the design of multimedia components that are functional according to the instantaneous demand of the system. Such programmable DSPs, controlled by a real-time operating system, are the wave of the future. Interactive Multimedia Association (IMA) and Microsoft recently announced preparation of a standard API for DSPs to further accelerate the DSP revolution, creating huge opportunities for development of many new specialized multimedia semiconductor and component devices.

DSPs are widely used in telecommunications equipment, including answering machines, cellular phones, and modems. The latest DSP products are designed as coprocessors for creating real-time multimedia capabilities that can be incorporated into motherboards of PCs and LAN-related hardware involved in multimedia traffic. Major multimedia growth areas for DSP applications include audio, codecs, MPEG and JPEG compression devices, and real-time multimedia.

AT&T Microelectronics manufactures the Hobbit DSP series, which is widely used in a range of multimedia hardware products. TI, jointly with IBM and Intermetrics, developed the MWave DSP, which includes an interface between the chip and audio, imaging, and telephone devices. Figure 16.1 illustrates some of these DSP architectures.

Audio DSPs

Audio DSPs employ dedicated algorithms for music or voice in specific add-on sound boards for PCs and workstations. This is the largest DSP multimedia

Traditional DSP architecture

Central memory architecture

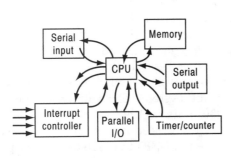

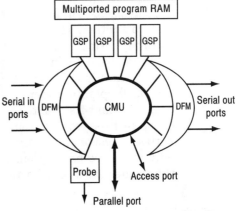

Central-processor-based

• CPU is a shared resource—must be
  involved in all system operations, limits
  the amount of processing, and complicates the
  software tasks.

• Processing and input/output can run
  simultaneously for higher throughput

Figure 16.1 Multimedia DSP architectures. (CMU = central memory unit that enables central memory architecture in stream processor DSP; DFM = data flow manager that allows multiple streams of data to be processed; GSP = circuit that carries out general signal processing functions.) (*Source: Star Semiconductor.*)

application market to date and will probably remain a major segment until multimedia DSPs are incorporated in motherboards of various platforms. Sales volume of these DSPs is expected to increase from 140,000 in 1993 to about 3,100,000 units in 1996.

Real-time multimedia DSPs

Real-time multimedia DSPs are the critical elements of all future multimedia hardware. These DSPs address the processing needs of all media such as fax, modem, voice, image, and video and with an appropriate real-time operating system can multitask between them. This means that a DSP-based multimedia board can perform JPEG decompression while doing data modem functions for electronic mail (E-mail) and processing an image or fax transmission. This multifunctionality is the driving force behind DSP-based multimedia products.

The architecture of multimedia DSP chips is optimized for heavy use of *multiply-accumulate operations,* which are predominant in compute-intensive compression algorithms. General-purpose DSPs cannot match the performance of specialized audio or video encoders but offer the advantage of multifunctionality, which is very important in the overall multimedia processing.

DSP chips have 50:1 to 100:1 advantage in terms of multiply-accumulate operations per second per dollar basis and are particularly attractive for multimedia applications. Using DSPs a CPU can manage various peripherals and a user interface while DSP handles real-time communications such as modem transmissions without being interrupted as the host system moves from one task to another. This means that integration of DSPs into multimedia hardware platforms facilitates the handling of audio and video compression and decompression during processing and reduces any delays in the process.

Basic multimedia DSP chips such as Texas Instruments MWave or AT&T DSP3210 operate under real-time operating systems as a partner with host processor and its standard operating system. The volume of such DSPs is expected to skyrocket from 200,000 units in 1993 to over 6,000,000 by 1996 with a corresponding decline in unit prices as economies of scale come into play with increased production. These are the DSPs that are being incorporated in the motherboards of PCs and workstations.

Codec DSPs

These DSP are the basis of most specialized audio and video coder-decoder circuits such as Intel's DVI and Indeo. They are also critical components of all videoconferencing systems and other real-time transmission hardware. Their growth is expected to be equally spectacular, from about 100,000 units in 1993 to 2,100,000 by 1996.

Compression device DSPs

These DSPs are dedicated MPEG and JPEG, as well H.261 and other standard compression-decompression chips for use in video and image playback and delivery systems. This is already a significant market segment with about 375,000 units shipped during 1993, increasing to over 3,000,000 units by 1996. Some of these functions are also being included in the basic real-time multimedia DSPs.

Interface Cards

Connecting the PC or workstation to the LAN involves a special interface which is known by various names such as *network interface unit* (NIU) or *terminal access point* (TAP), depending on the vendor. Interfaces differ according to access method used in putting the signals onto the LAN, which involves modulation and access protocols. Existence of an interface for LAN connectivity does not necessarily guarantee that it can be used for interactive multimedia traffic.

Baseband interfaces are inexpensive and require no special devices for generating the digital signals but allow only a single channel for communications. Broadband interfaces can generate signals at different frequencies and act, in effect, as modems but permit many communications channels. Access protocols define specific LANs such as Ethernet, token ring, or FDDI, and broadband interfaces usually support most major protocols automatically.

Interfaces are normally supplied by the LAN vendors and are a part of the package that includes cabling, software, and network management devices. LAN vendors usually offer a range of interface units from two to multiple connections and a racking system for stacking and connecting these devices. System integrators often undertake the provision of a total LAN solution which includes the installation and testing of all such components, and these items are basically invisible to most end users.

The trend for the future is to incorporate many communications functions on the motherboard of the PC or workstation, thus simplifying the connectivity and enhancing the performance of these devices by eliminating as much as possible any software processing. Under such circumstances it is necessary to identify access to a LAN only within the enterprise and configure the device to connect with it. What is important to keep in mind is whether compression schemes or codecs are included in such interfaces because these functions introduce significant latency into networking transmissions. While compression is a critical function for multimedia transmission, it is important to ascertain what particular algorithms are involved and make sure that the most efficient compression schemes are applied along all the networking paths that are likely to come into play.

Gateways

Gateways are interfaces designed to convert protocols, allowing devices that operate with one protocol to communicate with similar or other devices that use different protocols. As an example, some of the most common protocols in networking include TCP/IP, DECnet, and SNA. A gateway would convert a datastream from TCP/IP to SNA or vice versa, and so on. Gateways provide translation services between different protocols and allow devices on a network to communicate and not merely connect.

Gateways are very useful devices because they provide a LAN with extensive communication capabilities, particularly in internetworking environments. However, gateways are specific to particular applications, and since they perform a considerable amount of processing, they are relatively slower than bridges or routers. In effect, gateways provide the most intelligent, but slowest, connectivity between two LANs. This could also be a major disadvantage of gateways in networks that are expected to carry multimedia traffic where latency is a major issue.

Bridges

Bridges are communication hardware devices that connect or extend similar LANs as well as interconnect LANs to WANs. Bridges transfer data transparently and are protocol-independent. This in fact means that two LANs connected by a bridge appear as a single network system. Data transmitted across bridges is passed from point to point without any components being aware of the complete routing from origin to destination. If a destination is not recognized by a bridge, it sends the message across all possible links.

Bridges represent the simplest method of interconnecting two LANs to enable all stations of one to communicate with all stations of the other. Some bridges can connect devices of different protocols, but few provide flow control. Bridges operate at the data-link level, which deals with transmission of data between devices on the same network and provides some level of error detection without guaranteeing message delivery.

Bridges cannot tolerate network failures which severely impact bridge performance and can result in lost messages. As a result, bridges are less expensive than routers but can result in inefficient use of bandwidth. In a large network, nonuser data can use available bandwidth and degrade the response time. This means that bridges can become inefficient in large internetworks.

Because bridges perform little processing, they generally introduce the minimum transmission delays into the datastream. As a result, bridges may be more desirable in internetworking of LANs with multimedia traffic than other connectivity devices. Nevertheless, bridges linking two LANs with different bandwidths may present problems when data arrives at a bridge at a rate exceeding its processing speed. This can cause congestion of the bridge itself or of transmission links and the actual loss of part of the data. Hence bridges operate most efficiently when linking source and destination LANs that support identical message sizes and speeds. They are best used in departmental or small LAN environments.

Bridges are not addressable; they handle only sources and destination addresses, cannot access data envelopes, do not provide network conditions feedback, treat all packets identically, and provide only minimal data security. Bridges are available in three basic categories: the transparent, translating, and encapsulating bridges.

The transparent bridge

This type of bridge provides internetworking connectivity to LANs that employ identical protocols. Such bridges place the burden on communicating devices but need to acquire some knowledge of the location of devices on the LANs. The transparent bridge learns the addresses of each message, updates a forwarding table, and relays the message. If the bridge fails to find a match, it relays the message to all the known connections in a process known as "flooding."

The translating bridge

This is a specialized form of transparent bridge which provides interconnectivity to LANs with different protocols at the physical and data-link layers, as would be the case in linking Ethernet and token ring LANs. The translating bridge manipulates the envelopes associated with each type of LAN. Although processing performed by a translating bridge is relatively simple, the envelopes of different LANs vary in length, and the use of such a bridge requires that LAN stations and devices be configured to handle such variable traffic.

The encapsulating bridge

This type of bridge comes into play when there is a need to interconnect LANs through a backbone network which would generally be a higher-bandwidth FDDI facility. An encapsulating bridge provides connections to LANs that use identical physical and data-link protocols and are linked by a backbone with different protocols.

Unlike the translating bridge, which manipulates the actual message envelope, the encapsulating bridge places a message within a backbone-specific envelope, thus encapsulating it. In this form the message is transmitted to other bridges linked to the backbone, each of which removes the encapsulation, checks the destination, and either forwards it to a local address or ignores it. Once the message comes around to the originating bridge, it removes the encapsulation message from the backbone.

Clearly the encapsulating bridge performs more processing than do other types of bridge and creates more delay. This must be kept in mind when multimedia traffic is contemplated within such networks.

Routers

Routers are hardware devices that provide an intelligent link between networks and in fact extend the size of a network. They do this by examining network protocols and address information and then passing the data to an appropriate device in the other LANs. Routers develop a logical address partitioning and are able to determine where a device is located by address and then send data only to that address. As more sophisticated devices, routers are usually more expensive than bridges. Bridging routers are a special variety which support both routable and bridgeable protocols and offer the most flexibility in complex internetworking environments.

There are single and multiprotocol routers and, like bridges, they are not used to connect devices of two different protocols. Routers can support much more complex networks than can bridges and can be used to rearrange the architectures of LANs and WANs. Routers are also more difficult to engineer and maintain but offer more sophisticated and more complex service than those provided by bridges.

Routers are explicitly addressed, they access and use multiple sources of data, can open a data envelope and manipulate its contents, provide feedback on network status, forward envelopes to specific destinations, provide different types of service, and assure more complete security features in a network.

Where two users communicate through one or more intervening networks, the router actively selects the path between source and destination nodes taking into consideration cost of transmission, transit delay, network congestion, and physical distance between source and destination. These distances are usually measured in terms of the number of routers involved—"hop counts"—between the source and destination. Needless to say, these services come at a price, namely, processing delays that may be excessive for handling time-sensitive multimedia traffic.

A router processes only those messages that are specifically addressed to it by other devices. Router sophistication can certainly be used to select the maximum bandwidth for sending data, but it is also a complicating factor when different levels of internetworking traffic congestion exist within the networks. Routers communicate with each other to determine how to accomplish data transfer and which alternative route to choose, but this activity introduces considerable additional delays into the system. Because of these capabilities, routers are also more tolerant of network failures and can always find an alternate path to deliver the data.

As a result of performing this sophisticated network processing, routers introduce considerable transit delays that may become unpredictable as combined networks are loaded at different periods of time. Although in pure data transmissions such delays are often offset by higher availability of alternative routes throughout the internet, the overall latency involved is often too much for effective transmission of time-sensitive multimedia traffic. On the other hand, routers offer the opportunity to improve performance by segmenting a large internetwork into smaller networks and select only specific users in the whole system. As such, they may be useful in LAN segmentation for handling multimedia, although intelligent hubs are seen as a more appropriate approach because they reduce the latencies by eliminating many routers from the paths required for time-sensitive multimedia traffic.

Routers are available in three categories: standalone high-end proprietary hardware products, PC-based platforms, and software routers running on PCs that are low-cost devices suitable for connecting small remote LAN networks. In terms of interactive multimedia routers are a hindrance, but if they must be used, the high-end products whose processing throughputs are measured in the range of hundreds of megabits per second to 1 Gbps are obviously those that will be acceptable. Such throughputs are not uncommon among top hardware routers as well as increasing support for high-speed internetworking such as FDDI, ATM, SMDS, and frame relay in form of existing interfaces for these transmission technologies. Figure 16.2 illustrates use of routers in internetworking.

Switched Internetworks

Circuit switching transmits digital data, including voice and video transparently between two endpoints of a call. Since there is no attempt to interpret the meaning of the cell data, circuit switching is protocol-independent and can

In a campus network, linked by FDDI backbone, Cisco 4000 routers provide direct FDDI access for department groups, while high-end Cisco 7000 and AGS+ routers offer high-end routing among multiple FDDI rings.

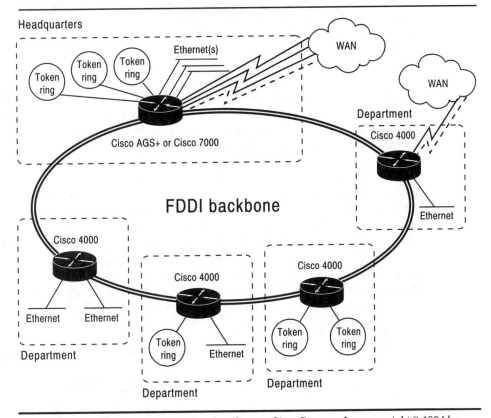

Figure 16.2 Use of routers in internetworks. (*Source: Cisco Systems, Inc., copyright © 1994 by Cisco Systems, Inc., all rights reserved.*)

transport large amounts of bandwidth with minimal delay. It is ideal for multimedia traffic.

As a result, perhaps the most significant development in both LAN and WAN networking is the availability of ATM switching technology, the basics of which are discussed in Chaps. 12 and 13. The most important aspect of ATM switching is the fact that it can be performed solely with hardware, eliminating the slower switching speeds which are the result of any hardware-software solutions.

Because of incompatibility of time-sensitive data such as voice and video with data-oriented LANs and their protocols, the entire industry based on router technology and products has come into being for interconnecting LANs and WANs. But ATM technology is designed to seamlessly and dynamically switch all kinds of network traffic between all users including voice, data, image, and video, obviating the need to use separate networks.

ATM switching allows users to mix and match broadband services such as mainframe, LAN, voice, and video traffic from private leased lines and public packet services and let them manage the data on public networks. The trend is toward switched internetworks which are developing on existing LANs and backbone routers that are in place but will enhance those internetworks with switching technology. Figure 16.3 illustrates the concept of a switched internetwork of this type.

The ATM switches now appearing on the market will interconnect departmental hubs and routers and form the basis of campus networks in the future. When carriers provide more widely varied high-speed services, these ATM switches can be used to link switched internetworks across a wide area.

Switched internetworks can accommodate multimedia applications much better than can any other networking solution such as Fast Ethernet or microsegmentation. These are interim solutions that do not address the basic issues, although they provide some protection for existing investments. Figure 16.4 illustrates an Ethernet switch solution which takes advantage of existing Ethernet infrastructure in a LAN but is clearly much more limited in scope than an ATM switch. Once switched internetworks have been developed for

Switched internetworks will build upon the LANs and backbone routers now in place. ATM-equipped smart hubs and routers will link shared LANs and establish connections to campus networks. New ATM switches will connect departmental hubs and routers and form the basis of campus networks.

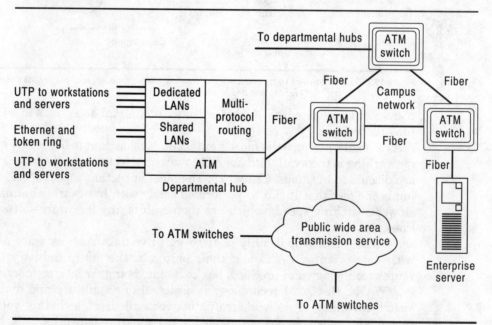

Figure 16.3 Switched internetworks concepts. [UTP = unshielded twisted pair (cable)]. (*Source: Reprinted from* Data Communications, *March 1993, p. 69, copyright by McGraw-Hill, Inc., all rights reserved.*)

Ethernet switches can deliver up to 10 Mbps per port—a much needed boost for overburdened networks. Best of all, they work with conventional Ethernet cabling and adapters.

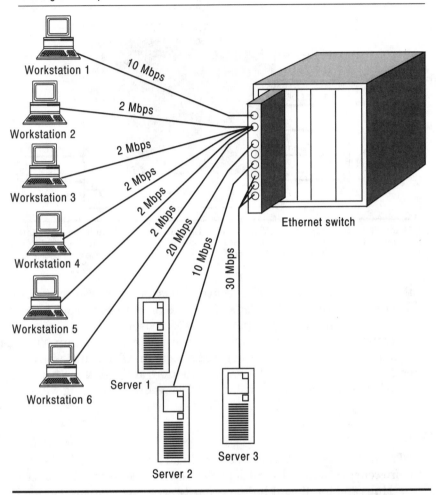

Figure 16.4 The Ethernet switch. (*Source: Reprinted from* Data Communications, *March 1994, p. 70, copyright by McGraw-Hill, Inc., all rights reserved.*)

enterprisewide client-server computing, the major obstacles to collaborative multimedia operations will be eliminated.

Switches coming to market differ from vendor to vendor, but all products use hardware-based cell switching within a chip or an ASIC circuit. As a result, switching of this type is much faster than routing or bridging. ATM switches also support multicasting, which is a crucial element for multiuser videoconferencing that forms the basis of collaborative multimedia. Another important attribute of switched internetworks is the ability to define restricted broadcast domains, which allows users to be assigned to any workgroup regardless of

ATM switches will be loaded with interfaces to the local and wide areas and will accept packets from conventional LANs and WANs, converting them into the cells used for ATM transmissions. Hardware-based switching at the chip level will make these boxes blazingly fast.

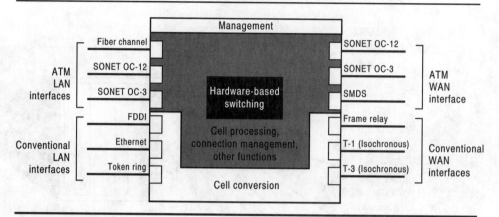

Figure 16.5 Anatomy of an ATM switch (SMDS = switched multimegabit data service). (*Source: Reprinted from* Data Communications, *March 1993, p. 76, copyright by McGraw-Hill, Inc., all rights reserved.*)

their physical location. This concept was previously discussed as the very desirable "virtual LAN" idea. Figure 16.5 represents conceptually an ATM switch with all the capabilities discussed above.

Routing is expected to remain and become integrated into ATM modules for connecting Ethernet and token ring LANs with ATM switches. Eventually when or if ATM is extended all the way to end users at the desktop networks, it will be based on end-to-end cell switching and conventional routing will not be needed.

Multimedia Intelligent Hubs

ATM switches are eventually seen as ideal multimedia hubs, but until that day, more conventional hubs are deployed to enhance the performance of LAN traffic. Hubs and intelligent hubs are hardware units that support wiring schemes and contain adapters that provide the Ethernet, token ring, or FDDI networking solutions. The majority of hubs include intelligence which provides control over all connections. LANs that use specialized servers and many protocols need a hub or concentrator to organize, consolidate, and assign priorities to individual end users. The concept of a hub with additional interfaces for multimedia traffic is illustrated in Fig. 16.6.

Switching hubs enhance throughput for LAN users through dynamic load balancing. A user who suddenly needs a large amount of bandwidth for multimedia applications could take over the entire bandwidth of a LAN. Switching hubs basically give a LAN user the ability to have a 10-Mbps connection directly to the server. Figure 16.7 illustrates a multimedia switching hub which provides access servers on an FDDI high-speed network. Eventually intelligent

Inside Lannet's LET-36 switching hub are seven buses, including a cell-switched 1.28 Gbps backplane and two token ring and four Ethernet buses. The vendor's 10Base-TV modules, which deliver video over 10Base-T LANs, plug into the 36-slot chassis and link to the high-speed backplane. Modules for token ring, Ethernet, FDDI, and Localtalk attach to the other buses.

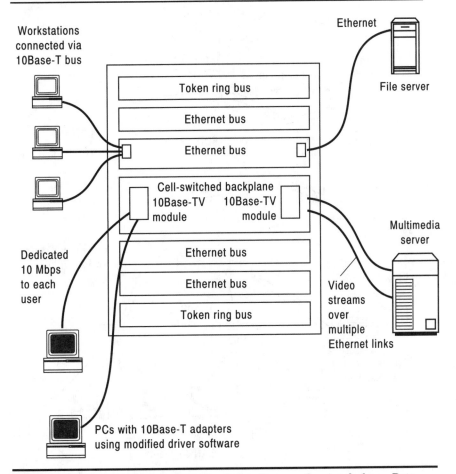

Figure 16.6 Anatomy of a multimedia hub. (*Source: Reprinted from* Data Communications, *January 1993, p. 80, copyright by McGraw-Hill, Inc., all rights reserved.*)

hubs will combine routing and high-speed switching in enterprise networks, particularly when migrating from shared to switched networks that are much more useful for collaborative multimedia traffic, as is envisaged for the interactive enterprise and the virtual corporation.

Videohubs, which link multipoint videoconferences using multiple codecs from various vendors and different switched digital circuits from multiple carriers, are also coming to market. These specialized intelligent hubs provide exactly the amount of bandwidth needed when required to each multipoint con-

The EIFO client-server switching hub connects up to 1024 clients on 10Base-T Ethernet LANs to several dedicated servers on FDDI LANs. EIFO has 12 10Base-T ports that can be used to attach servers on an FDDI LAN to a single high-performance workstation or to Ethernet LAN segments; hubs can be used to cascade multiple segments.

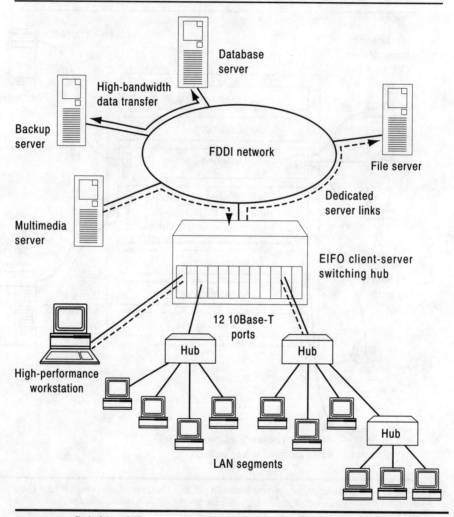

Figure 16.7 Switching hub access to FDDI servers. (*Source: Reprinted from* Data Communications, *August 1993, p. 45, copyright by McGraw-Hill, Inc., all rights reserved.*)

ference and simultaneous network access to MCU ports at any combination of standard transmission speeds.

Multipoint Control Units

Multipoint control units (MCUs) are complex hardware devices that automatically switch audio and video among all participants in a multisite videoconference. MCUs are a significant enabling technology that makes multimedia conferencing on a global basis possible by handling simultaneous transmissions from codecs of various vendors operating with different protocols. Codecs involved in a multisite videoconference send digitized audio, video, graphics, and data directly to the MCU, which transmits them to the appropriate participants. MCUs are discussed in more detail in Chap. 4.

New MCU hardware products address the problem of incompatibility of conventional MCUs and compliance with the ITU-TSS (formerly CCITT) H.243 multipoint conferencing standard, which recommends that MCUs should work with all codecs that conform to ITU specifications as well as with different carrier services. As collaborative multimedia conferencing becomes more popular, MCUs will play an increasingly important role within multimedia networks. More innovative MCUs can be expected on the market, as well as MCU functions within intelligent hubs, particularly as developments in DSPs continue to lower their prices and more multimedia data transmission standards are firmed up by ITU.

Major Equipment Vendors

Multimedia DSP manufacturers

DSP sales are now growing at over 30 percent annually, and the fastest segments are the multimedia-related DSPs. For many multimedia-related functions DSPs present the most cost-effective solutions, and in a year or two multimedia applications are expected to become the largest consumer of DSPs. Several major semiconductor manufacturers are involved, including Advanced Micro Devices, Analog Devices, AT&T Microelectronics, Hewlett-Packard, IBM, Intel, Motorola, National Semiconductor, and Texas Instruments. Among smaller ventures are firms like Cirrus Logic and C-Cube Microsystems. A number of vendors are also building various multimedia components based on DSPs. These include Best Data Products, Cardinal Technologies, Compression Labs, Digicom Systems, MediaStar, Octus, and Spectrum Signal Processing.

Major bridge vendors

Leading suppliers of high-performance bridges used in WAN, mainframe, and client-server connectivity include Cisco Systems, IBM, Madge Networks, Olicom USA, Proteon, Synoptics Communications, and 3Com.

Router manufacturers

Router market leaders include Cisco Systems, which controls 40 percent of the market, followed by Wellfleet Communications (17 percent), 3Com (9 percent), DEC (5 percent), Proteon (5 percent), IBM (5 percent), and Ungermann-Bass (3 percent). Most of their products support FDDI for internetworking connectivity.

Routers with ATM LAN interfaces are not generally available, but FiberComm, Netedge Systems, and Retix already manufacture products with such capabilities. ATM WAN interfaces are more readily available on routers from the above two vendors, as well as some products from Cisco Systems, Network Automation, Network Systems, and Plexcom.

Frame relay and SMDS support are more common on routers from Advanced Computer Communications, Andrew Corporation, Cisco Systems, DEC, Hewlett-Packard, Network Equipment Technology, Network Systems, and Wellfleet Communications.

Internetworking switch suppliers

ATM switches are being offered by a number of vendors, including ADC Fibermux, FiberComm, Fore Systems, GTE Government Systems, Network Equipment Technology, Network Systems, Newbridge Networks, Optical Data Systems, Synoptics Communications, and Whittaker Communications. ATM switches for private networks are available from Adaptive, Alcatel Data Networks, Ascom Timeplex, Cascade Communications, General Datacomm, Hughes Network Systems, Lightstream, MDR Teltech, Northern Telecom, and Stratacom.

Intelligent-hub suppliers

A number of vendors are offering smart hubs, with competition growing daily as it is considered a product line with a great future. Those suppliers include Alantec, Inc., Artel Communications, Chipcom, Kalpana, Lannet Data Communications, Sigma Networks System, Synernetics, and 3Com Corporation. ATM switches are also seen as ideal multimedia hubs, and Synoptics is already offering its ATM Latticell switch, which can act as an intelligent LAN hub, providing up to 5 Gbps of throughput to the users. Teleos Communications is a supplier that specializes in videohubs.

Multipoint control units

The first multiprotocol, multicodec MCU, came to market in 1992 developed by Videoserver, which since announced smaller versions of the products for small groups of participants. Competitive devices are being developed by BT North America and Digital Access.

Chapter 1

Building
Multimedia Networks

Traditional MIS managements regard multimedia as another information processing issue which involves large files and new data types. By the same token, network administrators see it as additional traffic on LANs and WANs whose bandwidths are already strained to capacity. In both instances technically they are quite correct, but more often than not they also fail to realize that multimedia is more than just a technical issue dealing with how to make a system work with a reasonable mean time between failures (MTBF).

There is an additional dimension to multimedia communications which is not technical in nature but usually must be resolved and implemented by technical managers and developers. This extra dimension is multimedia's potential to arrest and hold a user's attention and to motivate and excite viewers to action rather than to just provide data or information.

There are endless discussions about the value and purpose of multimedia in specific application areas, and one of the most common opinions or conclusions is that multimedia is a bunch of technologies looking for a market. This seems to be a very narrow-minded approach of information processing workers who have been weaned on the concepts that information processing applications and tools must be made to perform according to a variety of yardsticks such as rate of access to specific data, number of transactions per second, or bandwidth and delay in a network. These are all good and meaningful metrics, valuable in comparing one person's piece of software or hardware against another's, but they fail to take into account all the other factors that make up the "rating" of a presentation.

Most of the comparisons are also conducted on information processing procedures that are simply electronic representations of the established bureaucratic reporting and archiving processes and paper flows. The crucial aspect here is that the automation of data and information flows in most cases simply speeds up the generation of reports and floods everybody on the network with increasing amounts of data that are overwhelming in volume, extremely boring in nature, and tedious to follow and understand. Under such circumstances it is unlikely that most users will spot or take advantage of new business opportunities that could enhance productivity and competitiveness of their organizations.

What is being missed with regard to multimedia is the fact that first and foremost it has an extraordinary potential to enhance the user interface. No matter what is being displayed on the ubiquitous screen or transmitted through a network today, it can be done better with multimedia. There should be absolutely no question about it, and it is well to remember in this context the timeless cliche that a picture is worth a thousand words. Multimedia, which can be a moving picture with sound and interactivity thrown in, is clearly worth much more than that. What this simply means is that if you can demonstrate it, you will find it much easier to sell it, explain it, teach it, or even make a life-and-death decision about it. In short, overall productivity enhancement without meaning a spreadsheet, or a word processor.

Multimedia is all-pervasive, and the best proof of that is the fact that vendors of those same spreadsheets, word processors, and E-mail software are rushing in new versions with multimedia capabilities. Interactivity, which is inherent in all information processing today, means that sooner or later we all become involved with interfaces between the human being and the document, the system, or other human beings. Whether it's hypertext, a GUI front end, or a videoconferencing system, multimedia capability makes it not only more productive but more desirable and even pleasing to use.

This also means that whether it is a standalone or a networked multimedia interaction, it now comes under the rules of client presentation and customer entertainment. No matter how we want to look at it, just retrieving data or turning pages electronically is not going to suffice because it will be too easy for competition to steal clients away, just as is done every day with TV programs. Multimedia computing, whether we like it or not, will be judged under a different and more comprehensive set of rules. What's more, it will be judged not by just a few corporate managers but by the large client populations who will be directly exposed to interaction with whomever they want, particularly when the digital superhighway gets under way.

What this all means is that planning for multimedia must take a new dimension deeper and richer than planning for conventional information processing projects. This means new and unfamiliar skills and worker types that are incompatible with existing cultures and ways of doing business. It also means giving consideration to psychological, artistic, cultural, legal, and aesthetic factors that normally do not come into play in development of conventional information processing or networking projects. Above all, it means getting closer to the intended or targeted audiences and defining their specific needs, expectations, fears, prejudices, and dislikes before attempting to develop interactive multimedia communications. This is particularly true where the intended multimedia systems are designed to develop interactive enterprises and virtual corporation capabilities where unpredictability of potential audiences introduces even more stringent requirements.

This chapter is a special orientation piece designed to inform those who are involved with multimedia development and implementation about the special aspects that make it different from conventional information processing projects. The bulk of the discussion in this chapter deals with the various details involved with conceptualization, design, and implementation of multimedia applications and their impact on the user. In many instances interactive multimedia will consist simply of transmissions of applications or group conferencing where design of the screens is irrelevant as these reflect the actual real-time conditions which they represent. Nevertheless, the knowledge of multimedia design principles and issues is valuable in assessing multimedia applications and making adjustments in the user interfaces to obtain the greatest benefits.

Planning Multimedia Networks

In the changing corporate information systems environments of today there are several trends that impact the process of implementation of a multimedia network or perhaps more properly the introduction of multimedia capabilities into existing networking environments.

It is most unlikely that interactive enterprises and virtual corporations will be built from scratch by design. Rather, it will be an evolutionary process of expanding existing LANs and WANs to accommodate additional massive and time-sensitive multimedia traffic flows. Eventually as new networking technologies such as frame relay, SMDS, and ATM become widely available and prices come down, multimedia networking will become the norm rather than exception. Until that time, which incidentally is not too far into the future, enterprises or workgroups must decide whether there is an advantage in becoming an early adopter of multimedia networking based on their particular objectives, resources, and competitive environment.

The key issues that must be addressed before any planning begins relate to information technology and the relationship or status of a particular group or enterprise within that environment. The following is an outline of critical activities with that regard:

- Assessment of the current status of information technology in general and interactive multimedia communications in particular, with special focus on real-time collaborative conferencing technology

- Determination of multimedia technology development trends with special attention given to such factors as new product availability, obsolescence, and vendor support

- Evaluation of multimedia networking projects and applications that are under way within the industry of which the enterprise is a part

- Discovery and quantification of special business justification factors for introducing multimedia networking within the industry

- Comparison of the status of networking technology within the enterprise with that of competitive organizations

- Determination of the exact multimedia LAN or WAN technology that is appropriate for the enterprise during the next few years

- Definition of the resources required to implement such a multimedia network within a specific time frame

These issues must be addressed, keeping in mind that the most common business imperatives driving enterprise needs in the current era include rapidly intensifying global competition, restructuring for cost-effectiveness through consolidation of resources, continuous customer service requirements, emergence of client-server operating environments, and integration of text, data, voice, and video into multimedia information streams.

These factors also affect the traditional information system environments within enterprises with changing roles and relationships between the traditional entrenched MIS departments and new more volatile and autonomous end-user organizations.

The fact is that the old multiuser mainframe or minicomputer installations accessed with dumb terminals are being replaced by LAN-based PCs and workstations linked to servers, many of which are optimized to perform specific tasks such as, for example, videoservers in multimedia networks. Previous separate networks for specific applications with minimal connectivity are being phased out by shared-utility networks that can support interoperability between different systems and platforms. The development of proprietary software and interfaces has given way to multivendor environments and the use of standard operating platforms.

Of particular importance is the fact that while MIS in the old days was responsible for all applications and system development, currently it is more likely to deal primarily with corporate databases and networking infrastructures. Finally MIS backlogs are legendary, but in the present era there is a huge population of independent vendors who can implement software rapidly and directly for end users and a whole industry of system integrators who are expert in designing and building networks of all types. In brief, the traditional role of the MIS is being curtailed while that of the end user expands.

What this means is that the end-user group or groups within the enterprise are the more autonomous entities, but networking as such remains the domain of the shrinking MIS. At the same time, existing LANs and WANs are rapidly increasing their user populations with immediate effects on the performance and response of the networks.

All this is happening to networks transmitting mostly conventional text and data without much multimedia traffic, if any. This means that the priorities of MIS networking gatekeepers are to keep expanding the bandwidth and availability of enterprisewide networks to keep up with that demand. As a result, they are not overly receptive to propose and initiate multimedia networking and may even campaign against it on the grounds of lack of business justification and immature technology. This means that the initiative and the burden of proof must be on the end user, who must take action and acquire the necessary resources. Once these realities are understood, it is possible to define what is really needed and to engage in a multimedia planning process to meet that goal.

Identifying Target Audiences

Once the objective of the multimedia application is determined and its scope clearly defined, the next step is to analyze in detail the target audience of end users of the intended system. This is a particularly important step because the specific demographics of the user group will influence the content, format, and many other characteristics of the finished product. Too often the system is not well received or is seldom used because it reflects the preferences of the developers or a single creator with little or no input from the actual end users.

If the multimedia application is basically an enhancement of an existing information or expert system, the analysis of its current usage can yield valuable insights about end-user likes and dislikes and problems with the system. On the other hand, in cases like that, a multimedia interface may change the appearance of the system to the end user, but if the basic solution is not well designed in the first place, it will not necessarily improve the acceptance or usage very much. New concepts using multimedia elements from scratch designed to maximize end-user satisfaction are bound to prove more acceptable and productive.

Attention to the end-user demographics at the earliest stage is also dictated by the changing market conditions where customer needs and desires are becoming the decisive competitive factor. Customization of products and services is key to survival in the global marketplace, and since multimedia applications are very often designed in order to enhance sales and marketing functions, end-user concerns should be paramount.

Defining End-User Needs

Target audience appraisal must take into account the size of the user group. It is important to determine how many people will use the system and how they are dispersed because this will affect the budget directly. This question involves specific deployment expenditures for multimedia-enabled PCs, upgrading kits, adapters, routers, switches, hubs, and distribution costs, which may consist of mailing or downloading via modems or networks, publication of instructions, and service facilities to answer end-user questions and assist in solving operation problems. In cases of multimedia point-of-sale applications a whole set of procedures must be in place to service the customers effectively.

The size of the audience also affects the amount of time that will be required to test, evaluate, deploy, upgrade, and maintain the applications.

Another aspect of audience analysis is the frequency with which the system will be used. Some promotional multimedia presentations are used only once, introducing a product at a specific convention or conference. At the other extreme are the unattended networked merchandising kiosks in high-traffic areas in use on a 24-h basis with specific peaks in demand. In such cases it is necessary to develop simultaneous usage forecasts to design acceptable system response times under conditions of heaviest usage. This could be a limiting factor keeping in mind the necessity to transmit video, audio, and images that even in compressed format may present huge volumes of data.

Acquiring Intelligence on End-User Problems

Audience demographics play an even greater role in the conceptual design of a multimedia application. In order to create a successful solution it is necessary to take into account end-user motivations, age profiles, educational backgrounds, professional competence levels, and similar factors that will provide guidelines on how to design, structure, and present the multimedia content of the application. These inputs are critical to the integration of effective and appropriate content elements that will appeal to the end users and make them want to use the system.

The desires and preferences of the end users of multimedia applications must be seen as the most important criteria to be incorporated in conceptual and operational design. This may differ in many cases from typical attitudes of MIS organizations that often arrogantly believe that end users must be told what is right for them. Because of the critical role that content and format play in multimedia presentations, these attitudes must be put aside in the interests of developing more effective and meaningful multimedia applications.

There is a prevailing belief that multimedia applications that include images and video must approach broadcast TV quality for end-user acceptance. However, developers of multimedia applications should keep in mind that while quality is a function of content, it is also a function of the convenience of use and budgetary restrictions. Only close and continuous relationship with the end users in each particular case will provide true answers to these questions.

Required Network Characteristics

Once the needs analysis is completed, the user populations are known, and their specific needs are identified, it is possible to match those requirements with the characteristics of an existing network or develop specifications for a network that will meet these objectives. Whether it is a case of restructuring a LAN to carry multimedia traffic or building a new one where none existed before, the same analysis must be completed and evaluated.

These specifications can be listed as a series of performance characteristics which as a set will dictate the choice of an appropriate LAN or WAN as the case may be.

- Quantification of the immediate user population
- Long-term user population growth and provisions for any additional multimedia applications
- Multimedia traffic requirements for determination of bandwidth in terms of frequency of usage and peak demands
- Determination of priority criteria for interactive multimedia services including the following:

    Minimum throughput requirements
    Maximum tolerable latency
    Network efficiency
    Maximum acceptable cost per user
    Internetworking requirements
    Network security requirements
    Restrictions of services to specific groups

- Selection of an appropriate LAN or WAN technology to meet the performance criteria
- Assessment of network expansion potential and degradation levels with increased user population
- Migration path potential to higher-speed networks such as ATM in the future

Ideally, networks are sought to deliver very high throughput, with high efficiency and minimum transfer delay or latency. In real life most networks can optimize two out those three parameters but not all. In the case of time-sensitive multimedia traffic clearly one of the parameters that must be optimized is latency. The other two must then be chosen depending on specific applications.

One of the most sensible approaches to developing multimedia networking capability is to identify a small LAN segment with a justifiable business requirement to introduce interactive or collaborative multimedia features. Such a pilot project should be small enough to be easily manageable, and a lot of attention must be given to the users finding out what specific data types are needed, with what frequency, from what sources, and so on. Implementation of a small pilot of that type provides a learning platform for interactive multimedia technologies of the day. Most importantly, by the time such a pilot is completed, the technology will have advanced enough to allow introduction of multimedia in other areas where it previously might not have been cost-effective or practical.

Price-Value Relationships

The price to value ratio of an interactive multimedia application is strictly a function of the application which it supports. It is generally believed that desktop videoconferencing will be a major multimedia application because it can be justified on the basis of alternative expenditures for time and cost of travel. The same justification is applied to interactive corporate training applications. These are not time-sensitive in the same sense as desktop videoconferencing but can also be justified in terms of time and costs of travel to and from training centers. Studies of these multimedia applications have suggested that once these systems are deployed to about 200 users, the costs of system development are justified.

In the case of financial services a good case in point is the trading workstation which includes numerous sophisticated real-time data feeds, access to numerous databases, and value-added information networks as well as capabilities to distribute video streams from TV broadcasts or corporate training and information centers. In other industries such expensive and powerful workstations could not be justified but in trading, access to superior information and imaging can be related to large trading profits, only one of which could pay for the installation. Under the circumstances brokerage and trading organizations feel justified to invest in advance because they expect to match such facilities with traders who can exploit the capabilities to make a profit for all concerned. It seems quite probable that as time progresses and average PCs become much more powerful, multimedia trading workstations will be justifiable in many more organizations.

Cognition Issues

Several converging factors indicate that interactive multimedia will revolutionize the user interface. These are cognitive aspects dealing with such questions as to how much information human beings can absorb and use. Collectively these factors are bringing changes to the way information is acquired, processed, delivered, and shared by user groups working toward a common objective.

As a result of increasingly more powerful computers and widespread networking, organizations are experiencing an overkill of information, much of which is highly fragmented. It is not clear that such an information overload contributes to better decision making; in fact it often intensifies confusion.

Humans generally absorb more than 80 percent of their knowledge through sight and about 11 percent through hearing while the rest is acquired through smell, touch, and taste. There is a belief that the combination of all the senses results in experiences that are credited with better decision-making potential. Interactive multimedia communications with all its alternatives can create more experiential environments by combining information from various media into a single stream of knowledge. The interactivity associated with these applications provides a sense of involvement of several senses simultaneously and is believed to be more effective as a cognitive process.

Screen Design Issues

Although it is impossible to design every screen precisely in its final form, it is possible to define its function, level of interactivity, contents, and duration with some precision. This is best accomplished by a formal screen design document, which is a cross between an engineering detail drawing and a storyboard element.

The screen design document is the key to successful project management. It contains all the information required to construct the final screen and identifies the creators, media elements, sources, and decision makers involved in approval of the design. The document becomes of particular value when the inevitable changes and modifications are made because it reflects the reasons and the authority behind such actions.

At the very least the screen design document should identify each screen by name or number and indicate all other screens to which branching could occur under specific conditions. The content of a particular screen should be sketched in with enough detail for any independent developer to understand. In addition, the document should indicate such details as screen object and background colors, text size and fonts, type of graphics, animation, video overlays, and audio and music sequences, if any. Whenever possible, it should also indicate the existing or potential sources of all the elements required to design that screen.

Such documentation provides an excellent basis for resolving any future developer-user disputes. Although the preparation of the screen design documents adds to the time and cost of the project up front, this effort is more than compensated for when the development begins and inevitable changes and modifications start accumulating.

The screen design document is a critical tool for multimedia development management, but it must be used rigorously if it is to be of value. Strong management is more likely to produce acceptable results, even with mediocre resources, than poor management. Because of the diversity of skills involved and the nature of multimedia itself, management of these projects is an especially challenging task and good formal controls are extremely useful under those circumstances.

Graphics and colors

Here is an aspect of the screen design that is hardly considered in conventional information processing except for individual preferences that may have little to do with ergonomics or good design. When interactive multimedia is being used for prolonged periods of time, it becomes important what graphics and what colors are involved. Aside from the obvious need to design screens in such a way that color-blind workers can use them as readily as others, there are many aspects of color exposure and fatigue that are little known but that do affect the performance of personnel.

Psychological questions

Psychological issues may exist in various collaborative multimedia conferencing situations particularly when people are seen individually by the group and possibly other observers who are not known to them. On-line monitoring of marketing and customer service personnel is a standard procedure in many enterprises, but with a video monitoring system which videoconferencing really is there is a factor of apprehension that develops within people who must operate in such environments. If conferencing sessions are recorded for further use and this fact is known to the participants, this will affect their behaviors within the system. Rules of conduct must be established within such groups, and participants in these sessions may have to be allowed access to records of their own performance whenever they want. By the same token, security and priorities must exist to protect access by unauthorized parties to such archives. Another issue that may develop is the introduction into multimedia networks of outside materials for personal promotional and profit motives and the effect that such activities may have on other workers.

Aesthetic considerations

Successful multimedia systems depend on the merging of aesthetics and communications elements into engaging interactive presentations and not on the simple ability to demonstrate several media elements integrated into a single working system. High-level tools do not demand in-depth understanding of the underlying video, sound, and programming technologies. Creativity in design is much more important, and the direction of the overall development should be assigned to managers who appreciate creativity and are sensitive to related end-user issues and preferences rather than just technological performance.

Legal Issues

A multimedia aspect often ignored is the question of copyright and other associated legal issues that arise automatically any time a creative input is fixed in a tangible medium of expression. Individual media elements that are combined to make up the interactive multimedia application and the finished product itself are all subject to copyright protection.

Traditionally every time a new technology comes into play in the broader presentation industries such as films, TV, and cable, 10 to 15 years of litigation follow before legal precedents and parameters are fully established. Multimedia constitutes such a technological change, and many entities and institutions are waiting in the wings to start testing the legality of various aspects of multimedia production and delivery.

These legal aspects must be considered during needs analysis because there is always the potential that they will affect the budget and time required to deliver the application. In some cases it may prove to be more expedient to develop new materials rather than waste time and money waiting for permissions or face royalty payment and licensing fees whose level cannot be predicted in advance.

There is also the question of protecting the multimedia application itself against illegal use of its elements by third parties. This is important primarily in case of multimedia productions that are deployed in numerous locations for access and use by the general public. An associated issue is the question of network security and access control, but these are normally functions of the network operating systems.

In contrast to many conventional creative works subject to copyright protection, multimedia applications may be made up with elements from many authors who are not directly associated with each other. This raises the question of works made for hire and joint works because if the former is the case, the copyright belongs not to the author but to the author's employer or other third parties who may have paid for it. This means dealing with numerous parties before permissions are granted and royalty arrangements finalized.

A case of joint works, on the other hand, may result unintentionally from a collaborative effort with outside authors and consultants. Legally, each coauthor owns an undivided interest in the work and has the right to grant nonexclusive rights for its use without the consent of the other owners, although none may sell the work without the unanimous consent of all the authors involved.

These are only the very basic issues that must be faced before a project is undertaken unless it is performed by a single employee who can create all the content independently from corporate-owned resources. Otherwise costly and lengthy litigation is always a threat.

There are also other legal issues that must be kept in mind. These include trademarks, patents, privacy, libel and defamation, releases, and unions. The best advice is to get agreements in writing from all contributors during the planning stages and get those approved by the legal department before starting development and transmitting multimedia traffic over the networks.

Assembling Creative and Technical Resources

A major integration issue at the resources team level that is not yet fully understood by MIS managers is the question of creative artists, writers, and animators and their stand on quality of content. They are necessary to develop top-quality interactive multimedia applications and avoid the dreaded computer-based tedium that results when information technology types are the exclusive developers of these applications. Unfortunately creative people do not respond to managerial techniques in the same way as does the average corporate worker.

MIS management must learn to interact with such creative forces in order to develop competitive multimedia applications. They are facing different groups of people who are driven by incentives and factors other than those to which an average corporate worker will respond.

This merging together of the disparate types of people is the toughest integration of them all because it involves many intangibles that can be resolved only by sensitive handling of human personalities. The recommendation from those who have experienced many multimedia development situations is to make sure that managers accept first of all the need for creative people on their staff. Too often only lip service is given to creativity while budgets, deadlines, and administrative skills are sought and considered much more important.

This situation, if it persists, retards the introduction and acceptance of multimedia applications because lack of creativity will show quickly enough in form of sterile or dull presentations and a bored or sleeping audience.

By the same token, it is the creative types that will insist on nothing but broadcast TV quality or better when in fact something less could be quite acceptable to a particular target group.

All this is not to detract from tight administrative controls, particularly when such a multitude of hardware and software products are involved. The reality, however, is such that any multimedia implementation presents a large number of tradeoff alternatives between quality of the finished product and budgetary and administrative constraints. This is particularly so as CD-ROM recorders and software for manufacturing your own titles have now come down in price to below $10,000. MIS managers must accept and believe at the outset than in multimedia where content makes or breaks the application, quality of output is very critical and the creative people are the ones who can ensure it.

Overcoming Cultural Roadblocks

The best course of action is to admit the necessity of creativity and ensure that it is included within their output tempered by financial realities of the moment. Under no circumstances should creative ideas and quality advice of creative people be ignored despite majority support for such action from the other information technology workers within the project team.

Managers must realize that creative people are more willing to take risks and to experiment because only through that process can they come up with superior quality and uniqueness of product. This should be understood and allowed, but controlled to make sure that it does not turn into reckless behavior. They should also be aware that they will get the most out of their creative talent if there is no fear of criticism for mistakes that result from trying out new concepts.

The compromise that must be struck is to determine as soon as possible what is needed by the end user and document it as precisely as possible. This provides a basis to control the creative people who tend to impose their own visions on actual end-user needs. But at the same time, narrowing the scope of creativity should not appear as a form of punishment to creative people.

If creative people are badly handled and leave, this can jeopardize the project much more so than the departure of programming or accounting types.

Chances are their replacement will start the creative process all over again and have a different perception of what should be done. The objective is to keep creative people on the team for the duration. One good suggestion is to give them the responsibility for setting their own limits and their own schedules within financial and time constraints of the project.

Strategic Partnering

Because of the additional skills and external sources that are needed to develop effective multimedia applications, enterprises may not want to develop their own capabilities but rather form strategic partnerships with specialists and creators in particular fields. There are large numbers of very small consulting and multimedia developing firms that claim to provide various services associated with the creation, design, development, and implementation of multimedia applications. It is very difficult to assess such organizations even by inspecting their previous products because it is never certain to what degree such products represent their own creativity rather than that of their clients. One possible source of such advice will undoubtedly come from advertising agencies that are bound to become significantly involved in interactive multimedia advertising through private and public networks.

Licenses, Copyrights, and Permissions

When suitable art, graphics, animation, video, and audio materials exist within the corporation that could be useful in a multimedia application, this does not mean that rights and permissions for their use are automatic. Chances are that if any such materials have been developed by company employees or outside vendors on contract, ownership is vested in the company. Nevertheless, it is well worth the effort to check with the legal department to ensure that all materials of interest can be used freely and to obtain specific permissions and releases from those whose images, voice, or artwork may be used and displayed publicly or outside the company.

Content materials including illustrations, literature quotations, movies, video clips, documentaries, market data, music, and software are all protected by copyright. According to some attorneys, even some methods of presentation may be patentable.

Even previously used corporate materials that were purchased from outside sources may have restrictions on their use. Multimedia developers and users must realize that legal analysis of content materials is extremely critical to timely and cost-effective delivery of multimedia applications. It is important to keep in mind that such applications are much more visible than conventional information systems. As a result, there is a much greater chance that unauthorized use of intellectual property will be noticed, resulting in legal action that will delay the project or make its costs of production prohibitive.

Major Multimedia
Communications Vendors

This section provides an alphabetized list of all the vendors mentioned throughout the book. The reader ought to keep in mind that interactive multimedia communications is a new and highly fluid industry and the company names, addresses, and phone and fax numbers may change without notice. We believe that the list is current as of mid-1994, but new ventures are also being formed rapidly and there is a lot of merger and acquisition activity among companies in these categories.

Affinity Communications Company
7 Desoto Road
Essex, MA 01929
Tel: (508) 768-7480
Fax: (508) 768-7474

AimTech Corporation
20 Trafalgar Square
Nashua, NH 03063-1973
Tel: (603) 883-0220, (800) 289-2884
Fax: (603) 883-5582

Alantec
70 Plumeria Drive
San Jose, CA 95134
Tel: (408) 955-9000

Allen Communications
Wayside Plaza II
5225 Wiley Post Way, Suite 140
Salt Lake City, UT 84116
Tel: (801) 537-7805
Fax: (801) 537-7805

American Mobile Satellite Corporation
1150 Connecticut Avenue NW
Washington, DC 20036
Tel: (202) 331-5858
Fax: (202) 331-5861

Analog Devices
1 Technology Way
P.O. Box 9106
Norwood, MA 02062
Tel: (617) 329-4700
Fax: (617) 326-8703

Andersen Consulting
69 West Washington Street
Chicago, IL 60602
(312) 580-0069

Apple Computer, Inc.
20525 Mariani Avenue
Cupertino, CA 95014
Tel: (408) 974-6025

Applix, Inc.
112 Turnpike Road
Westborough, MA 01581
Tel: (800) 827-7549, (508) 870-0300
Fax: (508) 366-9313

Ardis Company
300 Knightsbridge Parkway
Lincolnshire, IL 60069
Tel: (708) 913-1215

Ascom Timeplex Inc.
400 Chestnut Ridge Road
Woodcliff Lake, NJ 07675
Tel: (800) 275-8550, (813) 530-9475,
(201) 391-1111

Ask*Me Multimedia, Inc.
7100 Northland Circle, Suite 401
Minneapolis, MN 55428
Tel: (612) 531-0603
Fax: (612) 531-0645

AST Research, Inc.
16215 Alton Parkway
P.O. Box 57005
Irvine, CA 92619-7005
Tel: (714) 727-4141, (800) 876-4278
Fax: (714) 727-9355

AT&T Global Business Video Services
Group
51 Peachtree Center Avenue
Atlanta, GA 30303
Tel: (800) 843-3646, (908) 658-6000

AT&T Microelectronics
555 Union Boulevard
Allentown, PA 18103
Tel: (800) 327-2447, (215) 439-6011
Fax: (215) 778-4106

Avid Technology, Inc.
One Metropolitan Park West
Tewksbury, MA 01876
Tel: (800) 949-AVID
Fax: (508) 851-0418

Banyan Systems Inc.
120 Flanders Road
P.O. Box 5013
Westboro, MA 01581-5013
Tel: (508) 898-1760
Fax: (508) 836-3277

Bell Atlantic
1320 North Courthouse Road
Arlington, VA 22201
Tel: (800) 442-0455

Best Data Products
9304 Deering Avenue
Chatsworth, CA 91311
Tel: (800) 632-2378, (818) 773-9600
Fax: (818) 773-9619

Beyond, Inc.
17 Northeast Executive Parkway
Burlington, MA 01808
Tel: (800) 845-8511, (617) 229-0006

Broadband Technologies
4024 Stirrup Creek Drive
Research Triangle Park, NC 27709-3737
Tel: (919) 544-0015
Fax: (919) 544-3459

Broderbund Software, Inc.
500 Redwood Boulevard
P.O. Box 6121
Novato, CA 94948-4560
Tel: (415) 382-4400
Fax: (415) 382-4582

BT North America
2560 North First Street
P.O. Box 49019
San Jose, CA 95161
Tel: (408) 922-0250, (800) 872-7654

Cabletron Systems
35 Industrial Way
Rochester, NH 03867-0505
Tel: (603) 337-2705, (603) 332-9400

CADAM, Inc.
1935 North Buena Vista Street
Burbank, CA 91504
Tel: (818) 841-9470

Cardinal Technologies
1827 Freedom Road
Lancaster, PA 17601
Tel: (800) 233-0187, (717) 293-3000
Fax: (717) 293-3055

CBT Systems USA, Ltd.
400 Oyster Point Boulevard
South San Francisco, CA 94080
Tel: (415) 737-9050
Fax: (415) 737-0377

C-Cube Microsystems
399-A West Trimble Road
San Jose, CA 95131
Tel: (408) 944-6300
Fax: (408) 944-6314

Cellular Data, Inc.
2860 West Bayshore Road
Palo Alto, CA 94303
Tel: (415) 856-9800
Fax: (415) 856-9888

Centigram Corporation
91 East Tasman Drive
San Jose, CA 95134
Tel: (408) 944-0250
Fax: (408) 942-3562

Chipcom Corporation
Southborough Office Park
118 Turnpike Road
Southborough, MA 01772
Tel: (508) 460-4900

CIMLINC, Inc.
1222 Hamilton Parkway
Itacsa, IL 60143-1138
Tel: (708) 250-0090
Fax: (708) 250-8513

Cirrus Logic, Inc.
3100 West Warren Avenue
Fremont, CA 94538
Tel: (510) 623-8300

Cisco Systems, Inc.
P.O. Box 3075
1525 O'Brien Drive
Menlo Park, CA 94025-1451
Tel: (415) 326-1941

Claris Corporation
5201 Patrick Henry Drive
Santa Clara, CA 95052
Tel: (408) 987-7000

Compression Labs, Inc.
2860 Junction Avenue
San Jose, CA 95134
Tel: (408) 435-3000
Fax: (408) 922-4608

Computer Teaching Corporation
1713 South State Street
Champaign, IL 61820
Tel: (217) 352-6363

Comsell Training, Inc.
500 Tech Parkway
Atlanta, GA 30313
Tel: (404) 872-2500

Concurrent Technology Developers
315 Mountain Highway
North Vancouver, BC V7J 2K, Canada
Tel: (604) 986-6121
Fax: (604) 980-7121

Connexus, Inc.
4131 North Central Expressway
Dallas, TX 75204
(214) 443-2600

Creative Labs, Inc.
1902 McCarthy Boulevard
Milpitas, CA 92035
Tel: (408) 428-6600

Datapoint Corporation
8400 Datapoint Drive
San Antonio, TX 78229
Tel: (210) 699-7000
Fax: (210) 699-7920

Dataware Technologies, Inc.
222 Third Street, Suite 3300
Cambridge, MA 02142
Tel: (617) 621-0820
Fax: (617) 621-0307

Digicom Systems, Inc.
188 Topaz Street
Milpitas, CA 95035
Tel: (408) 262-1277

Digital Equipment Corporation
146 Main Street
Maynard, MA 01754-2571
Tel: (800) DIGITAL, (603) 881-6155

Effective Communications Arts, Inc.
221 West 57th Street, 11th floor
New York, NY 10019
Tel: (212) 333-5656

Electronic Imagery, Inc.
1100 Park Central Boulevard, Suite 3400
Pompano Beach, FL 33064
Tel: (305) 968-7100

Enterprise Solutions, Ltd.
32603 Bowman Knoll Drive
Westlake Village, CA 91361
Tel: (818) 597-8943

Eon, Inc. (formerly TV Answer)
1941 Roland Clarke Place
Reston, VA 22091
Tel: (703) 715-8600

Excalibur Technologies Corporation
9255 Towne Centre Drive, 9th floor
San Diego, CA 92121
Tel: (619) 625-7900

Extron Electronics (RGB Systems)
13554 Larwin Circle
Santa Fe Springs, CA 90670
Tel: (800) 633-9876, (310) 802-8804
Fax: (310) 802-2741

FileNet Corporation
3565 Harbor Boulevard
Costa Mesa, CA 92626
Tel: (714) 966-3400

Fisher International Systems
Corporation
4073 Mercantile Avenue
Naples, FL 33942
Tel: (800) 237-4510, (813) 643-1500

Fore Systems, Inc.
174 Thorn Hill Road
Warrendale, PA 15086-7535
Tel: (412) 772-6600
Fax: (412) 772-6500

Frontier Technologies Corporation
10201 North Port Washington
Mequon, WI 53092
Tel: (414) 241-4555

Fujitsu Industry Networks
1266 East Main Street
Stamford, CT 06902
Tel: (203) 326-2700
Fax: (203) 326-2701

Future Labs, Inc.
19925 Stevens Creek Boulevard
Cupertino, CA 95014
Tel: (408) 973-7228
Fax: (408) 736-8030

Gain Technology
1870 Embarcadero Road
Palo Alto, CA 94303
Tel: (415) 813-1800

General Magic, Inc.
2465 Latham Street
Mountain View, CA 94040
Tel: (415) 965-0400, (415) 965-1830

Global Information Systems
Technology, Inc.
Trade Center South
100 Trade Center Drive
Champaign, IL 61820
Tel: (217) 352-1165

GPT Video Systems, Inc.
2975 Northwood Parkway
Norcross, GA 30071
Tel: (404) 263-4781

Grand Junction Networks, Inc.
4781 Bayside Parkway
Fremont, CA 94538
Tel: (510) 252-0726

GTE Corporation
1 Stamford Forum
Stamford, CT 06904
Tel: (203) 965-3533
Fax: (203) 965-2520

Hewlett-Packard
3000 Hanover Street
Palo Alto, CA 94304
Tel: (415) 857-1501, (800) 637-7740

Hitachi Computer Products (America)
3101 Tasman Drive
Santa Clara, CA 95054
Tel: (408) 986-9770

IBM Corporation
4111 Northside Parkway
Atlanta, GA 30327
Tel: (404) 238-2764
Fax: (404) 238-4302

IBM ImagePlus Systems
208 Harbor Drive
Stamford, CT 06904
Tel: (203) 973-5000

IEV International
3595 South 500 West
Salt Lake City, UT 84115
Tel: (800) 438-6161, (801) 466-9093
Fax: (801) 263-9980

Innosoft International, Inc.
250 West First Street, Suite 240
Claremont, CA 91711
Tel: (909) 624-7907

InSoft, Inc.
Executive Park West One, Suite 307
4718 Old Gettysburg Road
Mechanicsburg, PA 17055
Tel: (717) 730-9501
Fax: (717) 730-9504

Integrated Circuit Systems, Inc.
1271 Parkmoor Avenue
San Jose, CA 95126
Tel: (408) 297-1201
Fax: (408) 925-9460

Integrated Information Technology
2445 Mission College Boulevard
Santa Clara, CA 95054
Tel: (800) 832-0770, (408) 727-1885
Fax: (408) 980-0432

Intel Corporation
2200 Mission College Boulevard
Santa Clara, CA 95052-8119
Tel: (408) 765-1558

Interactive Networks, Inc.
1991 Landings Drive
Mountain View, CA 94043
Tel: (800) 468-8199, (415) 969-1000
Fax: (415) 960-3331

Kaleida Labs
1945 Charleston Road
Mountain View, CA 94043
Tel: (415) 966-0472
Fax: (415) 966-0400

Kalpana, Inc.
3100 Patrick Henry Drive
Santa Clara, CA 95054
Tel: (408) 749-1600

Kopin Corporation
695 Myles Standish Boulevard
Myles Standish Industrial Park
Taunton, MA 02780
Tel: (508) 824-9969
Fax: (508) 822-1391

Kurzweil Applied Intelligence, Inc.
411 Waverley Oaks Road
Waltham, MA 02154
Tel: (617) 893-5151
Fax: (617) 893-6525

Larse Corporation
4600 Patrick Henry Drive
Santa Clara, CA 95052
Tel: (408) 988-6600

Lenel Systems International, Inc.
19 Tobey Village Office Park
Pittsford, NY 14534
Tel: (716) 248-9720
Fax: (716) 248-9185

Lightstream Corporation
150 Cambridge Park Drive
Cambridge, MA 02140
Tel: (617) 873-6300
Fax: (508) 262-1111

Lotus Development Corporation
55 Cambridge Parkway
Cambridge, MA 02142
Tel: (800) 448-2500, (617) 577-8500

Macromedia, Inc.
600 Townsend Street
San Francisco, CA 94103
Tel: (415) 252-2000
Fax: (415) 626-0554

MCI Communications Corporation
1801 Pennsylvania Avenue NW
Washington, DC 20006
Tel: (800) 933-9029, (202) 872-1600
Fax: (202) 887-2443

MediaStar Corporation
14440 Cherry Lane Court, Suite 212
Laurel, MD 20707
Tel: (301) 206-9010

Media Vision
3185 Laurelview Court
Fremont, CA 94538
Tel: (510) 770-8600, (800) 348-7116
Fax: (510) 770-8648

Meridian Data, Inc.
5615 Scotts Valley Drive
Scotts Valley, CA 95066
Tel: (408) 438-3100

MFS Datanet, Inc.
55 South Market Street, Suite 1250
San Jose, CA 95113
Tel: (408) 975-2200

Microsoft Corporation
1 Microsoft Way
Redmont, WA 98052
Tel: (206) 882-8080
Fax: (206) 936-7329

Mitsubishi Electronics America
5665 Plaza Drive
P.O. Box 6007
Cypress, CA 90630
Tel: (714) 220-2500

Motorola, Inc.
1303 East Algonquin Road
Schaumburg, Il 60196
Tel: (708) 576-5000

Multimedia Learning, Inc.
5215 North O'Connor, Suite 200
Irving, TX 75039
Tel: (214) 869-8282

nCube, Inc.
919 East Hillsdale Boulevard
Foster City, CA 94404
Tel: (800) 654-2823, (415) 593-9000
Fax: (415) 508-5408

NEC America Inc.
1525 West Walnut Hill Lane
Irving, TX 75038
Tel: (800) 222-4NEC

Nikon Electronic Imaging
1300 Walt Whitman Drive
Melville, NY 11747
Tel: (516) 547-4200
Fax: (516) 547-0306

Netedge Systems, Inc.
P.O. Box 14993
Research Triangle Park, NC 27709-4993
Tel: (800) 638-3343

Newbridge Networks Corporation
P.O. Box 13600
600 March Road
Kanata, Ontario, Canada K2K 2E6
Tel: (613) 591-3600, (800) 343-3600
Fax: (613) 591-3680

New Media Graphics Corporation
780 Boston Road
Billerica, MA 01821
Tel: (508) 663-0666, (800) 288-2207
Fax: (508) 663-6678

Next Computer, Inc.
900 Chesapeake Drive
Redwood City, CA 94063
Tel: (415) 366-0900

Northern Telecom
3 Robert Speck Parkway
Mississauga, Ontario L4Z 3C8, Canada
Tel: (416) 897-9000

Novell, Inc.
122 East 1700 South
Provo, UT 84606
Tel: (800) 453-1267

Ntergaid, Inc.
2490 Black Rock Turnpike
Fairfield, CT 06430
Tel: (203) 380-1280, (800) 859-5218

Octel Communications Corporation
890 Tasman Drive
Milpitas, CA 95035
Tel: (408) 942-6500
Fax: (408) 942-6599

Octus, Inc.
9940 Barnes Canyon Road
San Diego, CA 92121
Tel: (619) 452-9400
Fax: (619) 452-2427

Opcom/VMX, Inc.
2115 O'Nel Drive
San Jose, CA 95131
Tel: (408) 441-1144
Fax: (408) 441-7026

Optibase, Inc.
7800 Deering Avenue
Canoga Park, CA 91304
Tel: (818) 719-6566, (800) 451-5101
Fax: (818) 712-0126

Optika Imaging Systems, Inc.
5755 Mark Dabling Boulevard
Suite 100
Colorado Springs, CO 80919
Tel: (719) 548-9800

Optivision, Inc.
 1477 Drew Avenue, Suite 102
 Davis, CA 95616
 Tel: (800) 562-8934, (916) 756-4429
 Fax: (916) 756-1309

Oracle Corporation
 500 Oracle Parkway
 Redwood Shores, CA 94065
 Tel: (800) 633-0583, (415) 506-7000
 Fax: (415) 506-7200

Panasonic Broadcast & TV Systems
 Two Panasonic Way
 Secaucus, NJ 07094
 Tel: (201) 348-7000
 Fax: (201) 392-4482

Performax
 3683 Post Road
 Southport, CT 06490
 Tel: (203) 254-1869

PictureTel Corporation
 One Corporation Way
 Peabody, MA 01960
 Tel: (508) 977-9500
 Fax: (508) 977-9481

PowerSoft
 70 Blanchard Road
 Burlington, MA 01803
 Tel: (800) 395-3525, (617) 229-2200
 Fax: (617) 272-2540

Projectavision, Inc.
 One Penn Plaza, Suite 2122
 New York, NY 10119
 Tel: (212) 971-3000

ProtoComm Corporation
 2 Neshaminy Interplex
 Trevose, PA 19053
 Tel: (215) 245-2040

Qualcomm, Inc.
 6455 Lusk Boulevard
 San Diego, CA 92121
 Tel: (619) 587-1121
 Fax: (619) 658-2110

Racotek Inc. (Raconet)
 7401 Metro Boulevard, Suite 500
 Minneapolis, MN 55439
 Tel: (612) 832-9800

RAM Mobile Data, Inc.
 745 Fifth Avenue, Suite 1900
 New York, NY 10151
 Tel: (212) 373-1930, (212) 303-7800
 Fax: (212) 308-5205

Scenario Systems
 3 Bridge Street
 Newton, MA 02158
 Tel: (617) 965-6458

SCO (The Santa Cruz Operation)
 400 Encinal Street
 Santa Cruz, CA 95061
 Tel: (408) 425-7222

ShareVision Technology, Inc.
 1901 McCarthy Boulevard
 Milpitas, CA 95035
 Tel: (408) 428-0330
 Fax: (408) 428-9871

Sierra On-Line, Inc.
 40033 Sierra Way
 Oakhurst, CA 93644
 Tel: (209) 683-4468
 Fax: (209) 683-3633

Silicon Graphics, Inc.
 2011 North Shoreline Boulevard
 Mountain View, CA 94043-1389
 Tel: (800) 800-4SGI, (415) 960-1980
 Fax: (415) 961-0595

Spectrum
 9 Oak Park Drive
 Bedford, MA 01730
 Tel: (800) 227-1127, (617) 271-0500
 Fax: (617) 275-5644

Spectrum Microsystems, Inc.
 320 Storke Road
 Goleta, CA 93117
 Tel: (805) 968-5100

Spectrum Signal Processing
 1500 West Park Drive
 Westborough, MA 01581
 Tel: (800) 323-1842, (508) 366-7355
 Fax: (508) 898-2772

Sprint Video
 P.O. Box 11315
 Kansas City, MO 64112
 Tel: (800) 669-1235, (913) 624-3000
 Fax: (913) 624-3281

SP Telecom
60 Spear Street
San Francisco, CA 94105
Tel: (800) 229-7782, (415) 905-4000

Starlight Networks, Inc.
325 East Middlefield Road
Mountain View, CA 94043
Tel: (415) 967-2774

Storm Technology, Inc.
1861 Landings Drive
Mountain View, CA 94043
Tel: (415) 691-6620, (415) 691-9825

StrataCom, Inc.
1400 Parkmoor Avenue
San Jose, CA 95126
Tel: (408) 294-7600

Structural Dynamics Research
Corporation, Inc.
2000 Eastman Drive
Milford, OH 45150-2789
Tel: (513) 576-2096
Fax: (513) 576-2135

SunSoft, Inc.
2550 Garcia Avenue
Mountain View, CA 94043
Tel: (415) 960-3200

SuperMac Technology
485 Portero Avenue
Sunnyvale, CA 94086
Tel: (800) 345-9777,
(408) 245-2202
Fax: (408) 735-7250

Sybase Corporation
6475 Christie Avenue
Emeryville, CA 94608
Tel: (510) 596-3500

Synernetics, Inc.
85 Rangeway Road
North Billerica, MA 01862
Tel: (508) 670-9009

Synesis Corporation
200 Hembree Circle Drive
Roswell, GA 30076
Tel: (404) 475-6788
Fax: (404) 442-0674

SynOptics Communications
4401 Great American Parkway
Santa Clara, CA 95054
Tel: (408) 988-2400
Fax: (408) 988-5525

Technology Applications Group, Inc.
1700 West Big Beaver Road
Troy, MI 48084
Tel: (313) 649-5200

Telco Systems, Inc.
Network Access Division
4305 Cushing Parkway
Fremont, CA 94538
Tel: (800) 776-8832, (510) 490-3111
Fax: (510) 656-3031

The Sierra Network
41486 Old Barn Way
Oakhurst, CA 93644
Tel: (209) 642-0700
Fax: (209) 642-0888

The 3DO Company
600 Galveston Drive
Redwood City, CA 94063
Tel: (415) 261-3000
Fax: (415) 261-3120

Time Warner Inc.
1271 Avenue of the Americas
New York, NY 10020
Tel: (212) 522-1626

United Medical Network Corporation
708 South 3rd Street, Suite 400
Minneapolis, MN 55415
Tel: (800) 991-4866, (612) 330-0990
Fax: (612) 330-0989

Verimation, Inc.
50 Tice Boulevard
Woodcliff Lake, NJ 07675
Tel: (800) 967-6366

Viacom International, Inc.
1515 Broadway
New York, NY 10036-5794
Tel: (212) 258-6508
Fax: (212) 258-6497

Videoconferencing Systems, Inc.
5801 Goshen Springs Road
Norcross, GA 30071
Tel: (404) 242-7566

VideoLabs, Inc.
5270 West 84th Street
Minneapolis, MN 55437
Tel: (612) 897-1995
Fax: (612) 897-3597

VideoLogic, Inc.
245 First Street
Cambridge, MA 02142
Tel: (617) 494-0530

Videoserver, Inc.
5 Forbes Road
Lexington, MA 02173
Tel: (617) 863-2300

Videotex Systems, Inc.
8499 Greenville Avenue
Dallas, TX 75081
Tel: (214) 234-1769, (800) 326-3576
Fax: (214) 994-6475

VTEL Corporation (formerly Video
Telecom)
1901 West Braker Lane
Austin, TX 78758
Tel: (800) 284-8871, (512) 834-9734
Fax: (512) 834-3794

Wang Laboratories, Inc.
One Industrial Avenue
Lowell, MA 01851
Tel: (508) 459-5000

Weingarten Publications
38 Chauncey Street
Boston, MA 02111
Tel: (617) 542-0146

Wellfleet Communications, Inc.
8 Federal Street
Billerica, MA 01821
Tel: (508) 670-8888
Fax: (508) 436-3658

Westbrook Technologies, Inc.
22 Pequot Park Road
P.O. Box 910
Westbrook, CT 06498
Tel: (800) WHY-FILE

Wicat
1875 South State Street
Orem, UT 84058
Tel: (801) 224-6400

WordPerfect Corporation
1555 North Technology Way
Orem, UT 84057-2399
Tel: (800) 451-5151, (801) 225-5000
Fax: (801) 222-5077

Workstation Technologies, Inc.
18010 Skypark Circle
Irvine, CA 92714
Tel: (714) 250-8983
Fax: (714) 250-8969

Xing Technology Corporation
1540 West Branch Street
Arroyo Grande, CA 93420
Tel: (800) 295-6458, (805) 473-0145
Fax: (805) 473-0147

Xyplex, Inc.
330 Codman Hill Road
Boxborough, MA 01719
Tel: (800) 338-5316, (508) 246-9900
Fax: (508) 264-9930

Glossary

ADSL Asymmetrical digital subscriber Line. This is a consumer telephone line whose bandwidth has been enhanced to T-1 levels with up to 18,000 ft of copper wire lines that will provide video-on-demand services.

A/D Analog-to-digital conversion. Usually refers to an A/D converter device which basically digitizes a continuous-waveform analog signal into a digital bit stream. It includes the steps of sampling and quantizing.

algorithm A sequence of processing steps performing a particular operation, such as compressing a digital image.

aliasing Undesirable visual effects in video usually caused by inadequate sampling. Jagged edges or curved object boundaries are the most common (See also **artifact.**) *Antialiasing* is software adjusting such effects.

animation Movement of an object on a screen from point to point (path) or displayed sequentially at specific time intervals (cycle).

ANSI American National Standards Institute.

API Application program interface. Formats of messages used to activate and interact with functions of another program.

APPN Advanced Peer-to-Peer Networking. A distributed client-server networking feature that IBM added to its Systems Network Architecture (SNA) designed to support efficient and transparent sharing of applications in a distributed computing environment.

artifact An unnatural or unintended object observed in reproduction of an image in a video system.

ASIC Application-specific integrated circuit. A semicustom chip used in a specific application that is designed by integrating standard cells from a library.

aspect ratio The relative horizontal and vertical spacing of pixels on a display screen.

asymmetrical system A video storage-and-display system which requires more devices and processing to compress and store than to play back an image.

asynchronous traffic A method of transmission which does not require a common clock but separates fields of data by stop and start bits.

ATM   Asynchronous transfer mode. A high-speed switching platform that can transmit voice, data, and video signals faster and more efficiently than traditional methods. It uses packets of fixed length of 53 bytes and is also known as BISDN or cell relay.

audio   Sound portion of a video signal, or separate sound used to annotate objects on frames such as text, graphics, animation, and still images.

audio buffer   Computer memory segment or separate device for storage of audio data for playback associated with an individual frame.

authoring language   High-level programming language using natural English or mnemonics specifically designed for developing multimedia applications.

authoring system   A software product designed for users without programming skills for developing and testing multimedia applications.

bandwidth   The range of frequencies that a given system is able to reproduce. The communications capacity of a transmission line or a specific path through a network measured in bits per second (bps). In a LAN, bandwidth is analogous to throughput.

baseband   A network where the bandwidth is taken up by a single digital signal such as in Ethernet and token ring LANs.

baud   A unit of speed defining the rate of transmission of binary data approximately equal to one bit per second (bps) at lower speeds. Common rates are 300, 1200, 2400, and 9600 bps available from common carriers.

BISDN   Broadband integrated services digital network. See also **ATM.**

bitmap   A sector of memory or storage which contains the pixels that represent an image arranged in the sequence in which they are scanned.

bitmapped graphic   A graphic image that can be accessed on a bit-by-bit (pixel) basis and is directly addressable on the screen.

BLOB   Binary large object. Defines very large datafiles such as are representative of multimedia including audio and video content.

bridge   A protocol-independent hardware device with an interface for connecting or extending LANs of the same type or connecting LANs and WANs. Bridges can be transparent, translating, or encapsulating types.

broadband   A network in which multiple signals can share the same bandwidth simultaneously through the use of multiplexing (splitting) the signal.

CAI   Computer-assisted instruction (see also **CBT**).

CBT   Computer-based training.

CCITT   Consultative Committee International Telephone and Telegraph. An international body which develops standards for voice and video transmission and compression over common-carrier and digital networks. (See also **TSS.**)

CDDI   Copper distributed data interface. An FDDI standard implemented on copper wires.

CD-I   Compact disk—interactive. A multimedia delivery standard introduced by Philips and Sony targeted at consumer and education markets.

CD-ROM   Compact-disk read-only memory. An optical disk storage device with 680-Mbyte capacity.

cell relay   See **ATM** and **BISDN.**

chroma keying   Facility to replace selected colors on a video image with others that allow creation of different scenes against the background. Also known as color key.

chrominance   Signals of an image that represent color components such as hue and saturation. A black-and-white (B&W) image has a chrominance value of zero. Also known as chroma.

circuit switching   A switching method in which a dedicated path is set up between the transmitter and the receiver. The connection is transparent because switches do not attempt to interpret the data.

client-server   An architecture that distributes computing responsibility between front-end and back-end programs. With two or more machines, client-servers can dramatically reduce network traffic and increase performance.

CMIP   Common Management Information Protocol. A protocol developed by IBM and 3Com and endorsed by ISO that provides a specification and formats for collecting network management data as an alternative to SNMP.

codec   Coder-decoder. A special processor that can digitize analog audio and video signals and decode digital data back into analog form.

compound document   A document composed of a variety of data types and formats each derived from application that created it.

compression   The translation of video, audio, or digital data singly or in combination to a more compact form for storage and transmission. Computer algorithms and other techniques are used to accomplish this compression process. (See also **JPEG, MPEG.**)

concurrence   Simultaneous transmission or occurrence of two or more events or activities within the same time period.

control track   Component of a video signal, exclusive of picture and sound, which provides essential synchronizing information.

cyberspace   Computer-generated electronic environment that is designed to give the user an artificial feeling of movement and discovery. (See also **virtual reality.**)

data rate   The speed of data transfer process normally expressed in bits per second (bps) or bytes per second (bytes/s).

DCT   Discrete cosine transform. A complex mathematical algorithm used in compression devices for eliminating redundant data in blocks of pixels on a screen. It is the basis for JPEG, MPEG, and CCITT compression standards.

digitizing   The process of converting analog electronic signals into digital format that can be stored, manipulated, and displayed by a computer. It is accomplished by special A/D converters in form of audio capture boards, video frame grabbers, scanners, or combinations of these in a single circuit board.

dissolve   Gradual fadeout of an image on the screen as another appears.

downlink   Earth station used to receive signals from satellites.

DSP   Digital signal processor. A specialized microchip designed to process efficiently digitized waveform data of sound and video. DSPs combine the high speed of a microcontroller with the numerical capabilities of an array processor.

DVI Digital video interactive. This is a compression format for recording digital video on a CD-ROM disk that provides up to 72 min of full-motion video, 4 h of $\frac{1}{4}$-screen full-motion video, or 14 h of $\frac{1}{8}$-screen full-motion video.

Ethernet A commonly used LAN protocol standard that allows network nodes to transmit packets at any time over coaxial, twisted-pair, and fiberoptic cabling. Packet collisions resulting from such transmission freedom can delay a packet during transmission.

fade A gradual change in brightness of an image or intensity of sound considered to be a special effect.

FDDI Fiber Distributed Data Interface. Standard developed by ANSI.

FDDI-Sync A variant of FDDI that provides priority to synchronous traffic on the LAN.

FDDI II A special new standard for isochronous LANs which carries traffic in channels instead of packets like FDDI.

flicker A phenomenon in a videodisk freeze frame when both video fields are not matched properly. A visible fluctuation of brightness of an image.

fps frames per second. See also **frame rate.**

fractal Fractional dimensional. Mathematical definition of a fractional element of an image after repeated application of a specific compression algorithm with theoretical compression ratio capability of 10,000:1.

frame A complete image in film or video consisting of two interlaced fields of 525 scan lines running at 30 frames per second (fps) in the NTSC system used in North America.

frame grabber A digitizer that converts video images into digital data that can be manipulated by a computer program.

frame rate Speed at which frames are displayed on the monitor. Standard broadcast TV rate is 30 frames per second (fps) in North America and 25 fps in Europe. Minimum acceptable movement frame rate is 15 fps.

FSIG FDDI Synchronous Implementers Group. Organization formed in December 1992 to expedite delivery of standardized distributed multimedia solutions to users.

full-motion video Display of video at the broadcast frame rate of 30 fps.

gateway interfaces designed to convert protocols between two different types of networks.

Gbps Gigabits per second.

genlock Synchronization generator lock. Permits combination of two or more video sources by synchronizing their signals together to produce a recordable composite video that can contain elements from each source.

GUI Graphical user interface. A screen interface with icons that allows direct manipulation of on-screen objects, menus, and dialog controls.

HDTV High-definition television. A TV screen with resolution comparable to movie theater or 35mm slide which requires at least 2 million pixels per frame. Standard NTSC TV resolution contains only 336,000 pixels.

headend Facility in cable system from which all signals originate. It picks up local and distant TV stations and satellite programming, and amplifies for retransmission through the system.

hypermedia Defines hypertext which contains a large percentage of multimedia content such as graphics, images, audio, and video.

hypertext Linked pieces of text joined together in nonsequential manner and accessible by navigation through a series of menus.

H.261 CCITT compression standard designed to facilitate the transmission of video images over digital networks at data rates ranging from 64 Kbps to 2.048 Mbps and also based on DCT algorithm. Also known as p*64-Kbps standard where $p = 1,2,3,...,30$ and intended primarily for videoconferencing and videotelephony.

IMA Interactive Multimedia Association. An umbrella organization grouping over 230 suppliers and end users to deal with multimedia standards and data exchange issues.

intelligent hub A hardware device that provides linkage between various LANs and automatically accounts for new topology as changes and expansion take place.

interactivity levels Pertains to interactive design features available with respective hardware configurations: level 1 = consumer devices; level 2 = industrial devices; level 3 = levels 1 and 2 devices interfaced with an external computer and peripherals. These do not relate to quality, values, or sophistication of contents and displays.

interframe coding A video compression technique which concentrates on coding high-detail areas of a picture.

intraframe coding A video compression technique in which half of the picture information is eliminated by discarding every other frame and displaying each frame for twice the normal duration during playback.

ISDN Integrated Services Digital Network. A set of digital network interface standards consisting of a signaling channel and a number of 64-Kbps digital transmission channels that are used to provide circuit-switched connections.

ISO International Standards Organization.

isochronous A communications capability that delivers a signal at a specified, defined rate, which is desirable for continuous data such as voice and full-motion video.

ITV Interactive television.

JPEG Joint Photographic Experts Group. A standard for compression algorithms for digitizing still images based on DCT with compression ratios ranging from 10:1 to 80:1.

Kbps Kilobits per second.

LAN Local area network.

latency The state of being latent; present but not visible or active. Applies to a power or quality that has not yet come into sight or action but may at any time. Used to describe the delay in network transmissions of data.

legacy system Mainframe- and minicomputer-based information systems that are critical to Fortune 1000 corporations in running day-to-day operations. About 50 to 80 percent of MIS (management information system) budgets are spent on maintenance of legacy systems.

lossless Any compression scheme that allows full recovery of original data.

lossy   A compression technique in which displayed and decompressed image does not contain all the original digitized data.

luminance   Brightness values of all points in an image.

MAN   *Metropolitan area network.* The name sometimes given to high-bandwidth networking facilities in a densely populated region such as a metropolitan area and its suburban business and institutional communities.

mastering   A real-time process in which videotaped materials are used to create a master optical disk which can be replicated into CD-ROMs.

Mbps   Megabits per second.

MCI   Media Control Interface. Platform-independent multimedia specification initiated by Microsoft in 1990 that provides a consistent way to control CD-ROM and video devices.

MCU   Multipoint control unit. A device for bridging three or more videoconferencing users of the same or differing protocols.

MHEG   Multimedia and Hypermedia Experts Group. An ISO activity concerned with coordinating specifications of multimedia design on any platform.

MIDI   Musical instrument digital interface. A series of digital bus standards for interfacing of digital musical instruments with computers.

MIPS   Million instructions (operations) per second.

MPC   Multimedia PC. Minimum multimedia hardware delivery platform standard MPC-1 and MPC-2.

MPEG   Motion Picture Experts Group. A standard for digital video compression which records only changes from frame to frame and is based on the DCT algorithm. Compression ratio is a tradeoff between motion video quality, size of window, and frame rate.

needs analysis   A critical phase of the interactive multimedia design process based on the needs of the end user.

NTSC   National Television Standards Committee. Defines North American color TV signal standards in 30 frames per second.

ODB   Object database. A database that can handle diverse and complex data including video, audio, bitmaps, graphics, animation, and unstructured text.

packet switching   Transfer of data by means of addressed packets or blocks of information. This differs from circuit switching because the network interprets some data and determines the routing during the transfer of a packet.

PAL   Phase alternation line. Color TV signal format in Europe and some other countries. Uses interlaced scheme with 25 frames per second and 625 lines per screen.

pixel   Picture element. The smallest element of a screen represented as a point of specific color and intensity level.

primitive   A basic element for display such as a point, arc, line, circle, alphanumeric character, or marker.

p*64   Same as H.261.

quantizing The process of converting analog values into digital with a limited number of bits.

RAM Random-access memory.

real time The transfer of data that returns results so rapidly in actual time that the process appears to be instantaneous to the user.

resolution A measure of image quality of a display. It refers to the number of pixels available on the display and controls the level of detail that can be presented.

RGB Red-green-blue. A color display signal consisting of separately controlled red, green, and blue beams which result in high-quality color output normally used in computer screens.

RIFF Resource Interchange File Format. Platform-independent multimedia specification developed by Microsoft in 1990 that allows audio, image, animation, and other multimedia elements to be stored in a common format.

RISC Reduced instruction set computer.

router Hardware device that provides intelligent links between networks. Routers recognize protocols and addresses and compute the most effective route for data transmission.

run-length coding A data compression technique that records repeated data elements with the same value only once along with a count of the number of times they occur.

sampling The process of reading and recording value levels of an analog signal at evenly spaced time intervals. It is a step in the process of digitization prior to encoding.

sampling rate The rate at which sampling occurs during digitization. Audio digitizing may involve sampling of 16 bits of data at rates as high as 48,000 times per second.

SECAM Sequential Couleur avec Moniteur. The standard for color TV developed in France and also used in Russia, eastern Europe, and some other countries. It compares with PAL at 25 frames per second, and the interlaced image is made up of 625 lines per frame.

SMPTE Society of Motion Pictures and Television Engineers. The SMPTE time code is standard 8-digit code used in identification of frames in form of HH:MM:SS:FF (hours, minutes, seconds, frame numbers).

SNA Systems Network Architecture. An IBM network strategy that defines communication methods for many IBM systems ranging from PCs to mainframes. Introduced in 1974, it is supported by many vendors.

SNMP Simple Network Management Protocol. A popular protocol that provides specifications and formats for collecting network management data as an alternative to CMIP. (See **CMIP.**)

SONET Synchronous Optical Network. A new standard for transmitting a variety of light signals over optical fibers allowing different fiber systems to interconnect efficiently with an unprecedented level of accuracy and customer control. It includes a hierarchy of transmission rates ranging from 51.5 Mbps up to 2.4 Gbps at present.

special effects Video image manipulation techniques for enhancing transition from one image to another or to create an unusual appearance. Includes effects such as dissolve, fade, and wipe, which are usually included as features in some videoboards and authoring systems.

storyboard Basic documentation of the proposed contents of a multimedia application, an advertising spot, or film. Prepared screen by screen and includes information about types of video, audio, and other objects that will be used. May also include details of navigational objects and interactivity levels at each point of the application.

synchronous traffic A type of transmission requiring that a clock signal be transmitted with the data so that both transmitter and receiver can agree on the time-related location of the bits. (Compare to **asynchronous traffic.**)

TCP/IP Transmission Control Protocol/Internet Protocol. A protocol developed by the U.S. Department of Defense to connect dissimilar systems on a network commonly used in UNIX networks.

timeout In an interactive system, the time limit within which a response must occur before a default branch of the program is executed automatically.

token ring A LAN protocol standard requiring network nodes to receive the circulating token—the permission to transmit—before the node can transmit a packet. It is commonly found in IBM's SNA environments.

touchscreen Pressure-sensitive display monitor that is often used as a multimedia control in lieu of or in conjunction with a keyboard. The most sophisticated touch screens also include z-axis control, which allows screen response at different rates depending on the level of pressure applied.

TSS Telecommunications Standardization Sector; previously known as CCITT.

tweening An animation technique where movement between key frames of a multimedia application is generated by the computer.

ULSI Ultra-large-scale integration. Generally applies to memory microchips with over 1-Mbit storage capacity and comparable levels of integration for microprocessors and other circuits.

uplink Earth station used for transmitting to satellites.

VCR Videocasette recorder. Can be used with an appropriate conversion device as input or output for multimedia applications.

videoconferencing Use of voice, with image, or video in communications with remote party over existing networks or public telephone links.

video dial tone New public services being introduced for transmitting video signals just like audio telephony.

videodisk An optical disk on which video signals have been recorded most often in NTSC analog format. Widely used as video input source for multimedia training applications.

Video for Windows Microsoft standard that allows end users to view video within a window of their screen.

video-on-demand Interactive TV concept for retrieval of video programs or movies from an interactive service at the consumer's convenience outside conventional schedules.

videotext Two-way interactive service using either cable or telephone links to connect a central computer to TV screens.

virtual Existing or resulting in an effect, although without factual basis. A virtual device may exist in a computer memory only representing a hardware peripheral.

virtual reality   Also known as artificial reality, cyberspace, and telepresence. It is the use of computers to simulate real environments with which a user can interact.

VTR   Videotape recorder.

VUI   Video user interface. A next-generation computing interface metaphor which will use a full-motion video window as part of the user interface and may employ icons to facilitate navigation.

WAN   Wide area network. A network connecting technologically incompatible devices or LANs over long distances and typically using common-carrier transmission facilities.

whiteboard   A feature of multimedia conferencing that allows users at various locations using pointing devices to simultaneously edit, draw on, and annotate documents that include word processing, spreadsheets, graphics, engineering drawings, and video.

wipe   A special effect in which one image pushes another off the screen. Many different approaches exist.

WYSIWYG   What you see is what you get. A user interface of many authoring systems where the author sees the screens while developing them exactly the way they will appear to the user.

Index

ABOUT THE AUTHOR

Bohdan O. Szuprowicz is president of 21st Century Research, a consulting firm engaged in continuous research of interactive multimedia communications markets and technologies. He is also the managing editor of monthly reports "Multimedia Intelligence" and "Interactive Information Technology." Previously, he acted as director of Interactive Multimedia Knowledge Systems on contract with Deloitte & Touche and was involved in promotion, marketing, and development of interactive knowledge systems for Computer Task Group. He consulted with many Fortune 500 corporations in system integration, computer services, financial services, pharmaceuticals, aerospace, and manufacturing industries. He is the author of "Multimedia Technology," "IBM Multimedia Strategy," "Multimedia Networks & Communications," and "Implementing Multimedia for Business" reports published by Computer Technology Research. He also authored hundreds of information technology–related articles published throughout the world and is a frequent speaker and panelist on interactive multimedia communications technologies. He can be reached at 21st Century Research, 41 Tappan Road, Norwood, New Jersey 07648-1708, USA; telephone: (201) 784-2475; fax: (201) 784-2480.